Sam Russell.

SNMP-Based
ATM Network Management

For a complete listing of the *Artech House Telecommunications Library*, turn to the back of this book.

SNMP-Based
ATM Network Management

Heng Pan

Artech House
Boston • London

Library of Congress Cataloging-in-Publication Data
Pan, Heng
     SNMP-based ATM network management / Heng Pan.
         p.   cm.—(Artech House telecommunications library)
     Includes bibliographical references and index.
     ISBN 0-89006-983-2 (alk. paper)
     1. Simple Network Management Protocol (Computer network protocol)
2 . Asynchronous transfer mode.   3. Computer networks—Management.
     I. Title.   II. Series
     TK5105.585.P36   1998
     005.7'1—dc21   98-28217
                                                                  CIP

British Library Cataloguing in Publication Data
Pan, Heng
     SNMP-based ATM network management. —(Artech House telecommunications
     library)
     1. Computer networks—Management   2. Asynchronous transfer mode
     I. Title
     621.3'8216

     ISBN 0-89006-983-2

Cover design by Lynda Fishbourne

© 1998 Artech House, Inc.
685 Canton Street
Norwood MA 02062

Portions Copyright 1994, The ATM Forum, Figures 3.18, 4.1, 5.1, 5.3, 5.5, 8.7

International Standard Book Number: 0-89006-983-2
Library of Congress Catalog Card Number: 98-28217

10 9 8 7 6 5 4 3 2 1

Contents

Acknowledgments

This book could not have been written without the help from a number of people.

First and foremost, I would like to extend my sincere gratitude to the reviewer, who gave generously of his time and expertise. This book has benefited from his valuable suggestions and comments.

I would also like to thank my wife and daughter for their support, understanding and patience during the later nights and weekends I spent preparing the book. Moreover, my wife prepared most of the figures for the book.

Many colleagues and friends of mine also contributed to this book in one way or another. Among them, Dr. Liqun Xu and Lloyd Wood reviewed parts of my draft. Dimitris Manikis and Geroges offered me the opportunity to access their ATM switch.

Finally, I would like to thank my editor, Susanna Taggart for her help and kind encouragement; and Dr. Julie Lancashire for offering me the opportunity to publish this book.

1

Introduction

Very few networking technologies have been embraced with as much enthusiasm as *asynchronous transfer mode* (ATM). Indeed, ATM is emerging as the most promising networking technology due to its speed, scalability, flexibility, and *quality of services* (QoS) guarantee.

ATM offers an ideal combination of circuit switching and packet switching techniques. Traditionally, most networks have been either circuit-oriented, for delivering delay-sensitive information like video or voice, or packet-oriented, for high-speed data transmission. Circuit switching guarantees end-to-end delivery and response times, but it can waste expensive bandwidth. Packet switching optimizes the use of bandwidth but has variable packet delivery times. Consequently, packet switching can be unsuitable for delay-sensitive traffic because the amount of delay between packets results in jittery transmissions. ATM combines the reliability of circuit switching with the efficiency of packet switching, making it possible to deliver deterministic QoS by using a single technology. This means that desktop video conferencing and distance learning applications can be smoothly delivered to the receiver.

ATM is scalable and provides greatly increased networking flexibility. Since there are no protocol differences between an ATM switch for the *local area network* (LAN) and the ATM switch used in a central office, the boundary between a LAN and a *wide area network* (WAN) is simply a matter of speed and cost, not a technical barrier. The uniform cell format used in both LAN and WAN environments will greatly simplify their interconnection. Through simplified packet-switching techniques, ATM segments packets into 53-byte cells, allowing for ATM switches to operate at up to gigabit-per-second transmission speeds.

1.1 ATM Is Coming

A key requirement for widespread ATM applications is the wide availability of
ATM end-to-end connectivity. Broadband networks are usually categorized
into backbone networks and access networks in terms of their respective roles.
ATM is already the consensus choice for multiservice backbone networks.
Major U.S. carriers, including AT&T, Bell Atlantic, MCI, Pacific Bell, and
Sprint, as well as many global carriers are deploying or have already deployed
ATM backbone networks. More than 40% [1] of the Internet backbone is
already being upgraded to ATM, and it is predicted that, over the next few
years, the Internet backbone will be completely ATM-based.

Access networks provide solutions for the "last miles" problem of broadband connectivity. There are two types of broadband access networks, cam-
pus/enterprise networks and residential networks. The former provides access
for business users, while the latter serves homes and small businesses. The
significant progress that has been made in both areas of broadband access
networks over the last few years will pave the way for large-scale ATM
applications.

In today's campus/enterprise networks, increasing numbers of users,
bandwidth-intensive applications, increasing Internet access through the LANs
in offices, and more workstations attaching to each network segment have all
contributed to less available bandwidth per station and inadequate performance
on some networks. Extensions of shared-media LANs and switched Ethernet
and token ring offer some tactical relief, but only ATM offers a fundamentally
different, strategic solution to the problem. Although it is still expensive, ATM
moves away from frame-based, shared-media LANs and multiprotocol enterprise networks toward a cell-based, switched environment that blends the LAN
and the WAN in a seamless, end-to-end protocol.

With a range of link speeds from 25 to 622 Mbps, ATM deployed in the
campus backbone offers unprecedented scalability for congested networks.
Since ATM networks permit devices running at different speeds to be mixed
together, network managers can selectively upgrade bandwidth-hungry devices
to the most appropriate speed, or add more ATM links to higher speeds.

In addition to scalability, ATM provides low and predictable end-to-end
latency. Unlike frame-based backbone networks, which can add variable
stored-and-forward delays at each hop, ATM's fixed-cell format delivers consis-
tently low switching latency. Moreover, ATM provides the mechanism to guarantee bandwidth and latency characteristics for applications that require QoS
support, enabling time-sensitive traffic such as voice and video streams along-
side burst, and delay-tolerant LAN traffic. Only ATM offers the scalability and
low latency to satisfy the ever-increasing bandwidth demands and long-term

and stringent QoS requirements needed for tomorrow's multimedia applications.

The *LAN emulation* (LANE) service defined by the ATM Forum provides a flexible mechanism for today's legacy networks to evolve toward the ATM networks of the future, without requiring the modification of existing protocols and applications. Existing applications and protocols can be adapted to ATM by merely installing new ATM interfaces and drivers. LANE supports backbone implementations such as directly attached ATM servers and hosts, as well as high-performance scalable computing work groups.

While LANE allows current applications to run seamlessly on ATM networks, the newly defined *multiprotocol over ATM* (MPOA) specification enables native ATM applications to make use of the true benefits of ATM technology, such as varying qualities of service, full ATM addressing, and other non-LAN protocol type applications. This provides an evolutionary solution, allowing existing applications to continue to work as before, while new applications are developed to take advantage of ATM.

Similarly, public telecommunications access networks are undergoing a rapid evolution to support new broadband services. In the residential sector, the days of limitation to telephony and broadcast television are coming to an end. There are signs that interactive broadband services to the home are becoming a major growth industry. This is being accelerated by competition between telecommunications companies and *cable television* (CATV) operators, as the telecommunications market becomes liberalized.

Cable operators are investing in their *hybrid fiber/coax* (HFC) distribution infrastructures to take advantage of new opportunities created by cable modem-based services. ATM has been identified as the only technology that works with the existing HFC infrastructure and enables full services, including high-speed cable modem services, *plain old telephone service* (POTS), advanced telephony services, Interactive digital video services, and Internet, *World Wide Web* (WWW), and communication services.

ATM is an excellent technology of choice in the backbone network. Accordingly, by utilizing ATM in the HFC distribution network, a cable operator improves end-to-end network performance, eliminates the complexities of interworking different data transport protocols, and simplifies network management. An ATM-based distribution infrastructure can support a variety of in-home systems, allowing greater flexibility today and a migration path for tomorrow. Furthermore, an ATM-based HFC network gives interactive broadband services providers an effective Internet communications service to the home, greater network efficiency and flexibility, and complete computing and communications capabilities.

Scalability is a key requirement of the HFC network. ATM is scalable in every key aspect, providing simultaneous high-performance connections at speeds ranging from T1 to OC-12c (1.5 622 Mbps). As the number of subscribed users and/or services increase, telecommunications providers can extend the ATM network with higher speed facilities without changing the network topology. Moreover, ATM's scalability means that the network can continue to grow without replacing the switches upon which it is built.

As cable companies deploy ATM or packet-based cable modem solutions, local telecommunications firms are preparing their own *to-the-home* solutions, based on *asymmetric digital subscriber line* (ADSL) over twisted pair. According to recent specifications, these ADSL systems are also utilizing ATM, with advantages similar to those offered by ATM-based HFC networks.

It should be noted that the full benefit of ATM can only be exploited when multiple services are transmitted over the same network. Although ATM is capable of integrating existing services with a unified technology, its performance is unfortunately optimized for neither pure voice applications nor pure data applications. For example, fast Ethernet has proved to be more economical than ATM in delivering data services.

1.2 Challenges of ATM Network Management

All roads lead to ATM. However, this comes at a price. Because of its inherent sophistication and flexibility, ATM is a very complex technology, perhaps the most complex one ever developed by the network industry. The deployment of ATM networks requires the overlay of a highly complex, software-intensive protocol infrastructure. This also leads to an unprecedented level of complexity in the management of ATM networks. Solutions that were adequate for other network management technologies are insufficient for ATM, particularly in mission-critical applications. Accordingly, effective ATM management must address the unique features of ATM, such as a connection-oriented environment, varying classes of service, higher volumes of traffic, virtual network configurations, and multiple traffic types.

ATM networks operate at extremely high speeds. The management of an ATM device, especially an ATM switch, may be very resource-consuming. For example, collecting performance statistics can be a computational intensive task. Software and hardware functionality in an ATM device must be carefully partitioned to allow high-speed processing. In addition, because of the expense of high-speed networks and the fact that in an ATM environment corporate information flow may be aggregated over one or a few integrated broadband links, effective network management becomes very critical.

ATM is a scalable technology, supporting an almost unlimited number of virtual connections at high speed. A single fault in an ATM network may result in a plethora of alarm messages, making it virtually impossible for a network administrator to properly read, analyze, and act on each one. This requires an intelligent, proactive approach to event filtering and correlation and to performance tuning processes both in ATM devices and across a heterogeneous network. The mechanisms should ensure that trouble tickets are intelligently generated for relevant alarms and events only, rather than for insignificant events, and that only one ticket is open for each problem.

Since ATM is connection-oriented, virtual connections must be managed effectively. As a result, end-to-end management is necessary in an ATM network. A mechanism should be available that allows for the sharing of configuration management data between various network management applications. The network management system must allow for the establishment of *permanent virtual connections* (PVCs) in response to user service requests and according to QoS parameters such as throughput and delay. For quick and responsive network management, the configuration capabilities should also allow for easy browsing, changing, and the automatic deletion of PVCs using expert system rules and procedures or through a *graphical user interface* (GUI).

ATM networks provide guaranteed QoS levels for multiple traffic types such as voice, video, and data; this introduces a new complexity. Traffic management plays on an increasingly important role of the network management system. Guaranteeing the desired service level through configuring connection parameters, setting traffic priorities, and controlling high-volume and high-speed traffic flows becomes an important priority.

While ATM provides sophisticated services to end users, hackers and competitors can also exploit it if proper security measures are not implemented. ATM virtual circuits and high speeds pose new security issues. For instance, because of the connection-oriented nature, it is very difficult to implement firewalls, which are used extensively in today's Internet. The security management capabilities of existing tools are inadequate to address the many ATM security issues.

ATM also warrants new accounting and charge-back mechanisms [2]. The different service classes and QoS attributes make accounting management very complex. To help control resource allocation and consumption, accounting and billing mechanisms should be instituted through the network management system. This helps ensure that costs remain within budget and that necessary costs are recouped. Account management is an emerging requirement for ATM networks.

ATM network management should also be capable of supporting the management of applications running over ATM networks, permitting seamless integration of network, system, and application management. Furthermore, it must be standard-compliant and inter-operable in a multivendor environment.

The problem of ATM management is further compounded by the clash of two cultures from the formerly almost independently developed networking industries, telecommunications and data communications, since ATM is the technology where these two industries converge. There has been a long-standing war between the *open system interconnection* (OSI) network management framework standardized by the *International Standards Organization* (ISO) and the Internet management framework. The former is adopted by telecommunications providers in their *telecommunications management network* (TMN), whereas the latter, which is denoted by the protocol used, namely the *simple network management protocol* (SNMP), is the choice of the data communications industry.

This book focuses on SNMP-based ATM network management, specifically configuration, performance, and fault management. Apart from some research projects, the first commercially available ATM switches are almost exclusively managed by the SNMP. Compared to the OSI network management framework, which has never gained sufficient support outside TMN, SNMP has become the management framework of choice and is now in pervasive and continuous use. Many *management information base* (MIB) documents have been produced for a plethora of uses in network management, system management, application management, and proxy management of legacy devices. As such, SNMP offers seamless integration with applications running over ATM networks, taking into consideration that most ATM applications today and in the foreseeable future are likely to be Internet-oriented. Tremendous efforts have been devoted to the standardization of SNMP managed objects in managing almost every aspect of ATM, offering an unprecedented level of manageability, which is off-the-shelf and interoperable. As a matter of fact, SNMP is the dominant platform for today's ATM management.

1.3 Organization of the Text

This book offers a comprehensive introduction to the SNMP-based ATM network management, thereby providing readers with an understanding of today's most extensively used ATM network management technology and a thorough grasp of relevant existing and evolving standards. Accordingly, the book discusses the architecture of the SNMP-based ATM network management as well as the mechanics of the management of ATM networks and various applications running over ATM.

The book consists of three parts. Part One is an overview of the ATM network and the SNMP protocol that helps readers to review or learn the fundamentals of both areas. Chapter 2 introduces the fundamentals of the SNMP framework. Chapter 3 discusses the background of ATM technology. Chapter 4 examines ITU-T and the ATM Forum's models on ATM management and provides a brief comparison of SNMP and TMN.

Part Two covers standard MIB modules for ATM management. Chapters 5 and 6 detail the management of ATM, including the ATM Forum's *integrated local management interface* (ILMI) MIB and the AToM MIB from IETF. Chapter 7 examines the IETF *synchronous optical network* (SONET)/SDH MIB as an example of physical layer management. Chapters 8 and 9 are devoted to application management including ATM Forum's LANE and *private network-network interface* (PNNI), respectively. Within each chapter of Part Two, materials are arranged in such a way that the operation of the protocol is explained before the relevant MIB is introduced so that readers overcome the inherent difficulties of understanding MIB objects.

Part Three is an example demonstrating how various standard MIBs and proprietary MIBs are used together to manage a real ATM switch. Chapter 10 describes the roles that proprietary MIBs play within the context of SNMP framework. Chapter 11 examines how AToM MIB is complemented with an extension from Cisco to realize full management of ATM virtual connections. Chapter 12 is a case study in which a Cisco MIB improves traffic management. Chapter 13 investigates the support of *operation, administration, and maintenance* (OAM) flows by a proprietary MIB, which does not exist in any standard MIB at all. The book concludes with a summary of the latest developments in ATM management.

1.4 Intended Audience

This book is written for a broad range of readers who are interested in ATM network design and management, including the following:

- *Students and professionals in telecommunications and data communications:* This book is intended as a basic tutorial and reference source for this exciting area.
- *ATM network designers and implementors:* This book discusses one of the essential aspects of designing an effective ATM network.
- *Network management system designers and system managers:* This book provides an understanding of the functions needed in an ATM

network management facility and offers information on current and evolving standards.

Since this book is almost self-contained, readers only need a basic understanding of Abstract Syntax Notation One (ASN.1) and networking.

In order for readers to stay informed about the latest development in related areas, as well as to get useful tools and documents when reading this book, a special web page dedicated to the book can be found at the related link "Link to Network Management Sites" on the following web site:

http://www.artech-house.com/

References

[1]   Fore Systems, "ATM and the Hybrid Fibre/Coax Cable Distribution Infrastructure," http://www.fore.com/atm-edu/whitep/index.html, White Paper, October 1996.

[2]   Stanford Telecom, "Network Management for ATM," http://www.stel.com/stel/atmps/atmps.htm, White Paper, 1996.

2

Simple Network Management Protocol

2.1 Introduction

The Internet-standard network management framework, also known by the protocol it uses, SNMP, has achieved unprecedented success in providing interoperable solutions to the problem of managing networks, enabling effective monitoring and control of heterogeneous devices. Today, SNMP is the most popular network management framework. Throughout this book, the term SNMP refers to the framework rather than the SNMP protocol only, unless otherwise explicitly indicated.

As its name suggests, the philosophy behind SNMP is indeed simple. The impact of adding network management to managed devices should be minimal, reflecting a lowest common denominator. As a result, an SNMP agent can be implemented even in the smallest and cheapest of systems, and the operational role and performance of a system is not compromised by the inclusion of an SNMP agent.

Currently, there are two versions of the SNMP management framework. The first version, referred to as SNMPv1, consists of three documents: RFC 1155 [1], RFC 1212 [2], and RFC 1157 [3].

Although it was originally designed to accommodate the short-term needs of the network vendor and operations communities, SNMP-based network management has been accepted as the Internet-standard management framework. In the process, it has become deeply entrenched as the management framework of choice, and it is now in pervasive and continuous use. Many MIB modules, which are virtual information stores containing managed objects, have been produced for a plethora of uses in network management, systems management, applications management, proxy management of legacy devices, and manager-to-manager communications. These MIBs reach virtually every network technology. Now SNMP is taking over additional market segments while increasing its strength in the areas of telecommunications management, data and voice over cable, systems management, and applications management.

SNMPv1 was satisfactory but far from perfect. Its most serious weakness is its lack of security. Consequently, while SNMP has seen tremendous growth in popularity for monitoring networks, it has not been so widely used for controlling networks. Another major weakness is that its performance in retrieving large amounts of data is poor.

To address the above-mentioned weaknesses and other deficiencies, party-based SNMPv2 (RFC 1441-1452) was published in early 1993. An SNMP party can be thought of as the identity of a particular protocol entity running at a particular network location and in a particular security context. A particular protocol entity may operate as any of several parties, but each party uniquely identifies that protocol entity. For example, one party uses no authentication and no privacy, and another one uses both. This framework provides industrial-strength commercial grade security, including authentication and privacy, and support for contexts. In addition, it improves efficiency and performance through such features as a new bulk retrieval mechanism.

While implementable and interoperable, party-based SNMPv2 is unfortunately not deployable, as it is too difficult to configure and use. As a result, only very few technology providers could develop correct SNMPv2 implementations, and even fewer organizations could actually deploy them. SNMPv2 is also too expensive in terms of agent resources. Thus, it has gained little vendor support and has never been accepted by the market.

To provide a workable SNMPv2 framework, a new set of RFCs (RFC 1902-1908) was published as draft standards. This version retains all the attractive features of party-based SNMP except its security, administrative framework, and remote configuration MIB and small parts of the protocol operations document.

A detailed description of SNMP is outside the scope of this book. This chapter aims to describe the fundamentals of SNMP, including the management model, the components of the SNMP framework, and Internet-standard MIB. The goal is for readers to gain enough background to understand the MIBs that will be introduced in later chapters and the relationship between Internet-standard MIB and other MIBs.

The chapter's discussion will focus on the current version of the latest SNMPv2 framework as specified in RFC 1902-1908 [4, 10]. Fortunately, the differences between various versions of the SNMP framework have little impact on MIBs. For a detailed list of relevant RFCs, please refer to Appendix A.

2.2 Management Model

The SNMP management framework includes four components:

- *Managed nodes,* each containing a processing entity termed an *agent;*
- The *network management station*—also known as the *manager*—on which management applications reside;
- A *network management protocol* that is used to exchange management information between the management station and the agents;
- *Management information,* which is specified by the *structure of management information* (SMI) and the definition of managed objects, i.e., MIBs.

2.2.1 Management Station

Management stations execute management applications that monitor and control managed devices. They are typically stand-alone devices but may also be processing entities that are implemented on a shared system. The SNMP protocol is used to convey management information between a management agent and a management station. A management station generates SNMP requests to request management information from an agent and responds to traps or inform-notifications received. Management applications usually provide a GUI, allowing human operators to interact with a remote MIB module. They may also have routines for such functions as data analysis and fault recovery.

2.2.2 Management Agent

The management agent is a processing entity within a managed element that is responsible for performing the network management functions requested by the network management stations. Managed elements are devices such as hosts, gateways, and terminal servers. Every agent has access to management instrumentation and keeps a collection of its managed objects residing in a virtual information store, i.e., the MIB. The agent responds to requests for information and action from a management station and transfers important but unsolicited information to the management station via the SNMP protocol.

2.2.3 Proxy

A proxy agent is an SNMP entity that appears to be acting in an agent role. Instead, however, it satisfies management requests by acting in a manager role with a remote entity.

The proxy agent permits the monitoring and control of managed elements that are otherwise not addressable using the management protocol and the transport protocol. In this scenario, the proxy agent acts as a proxy for one or more non-SNMP devices; that is, the SNMP agent acts on behalf of the proxied devices. It may provide a protocol conversion function, allowing a management station to apply a consistent management framework to all network elements, including devices that support different management frameworks.

A proxy may potentially shield network elements from elaborate access control policies. For example, a proxy agent may implement sophisticated access control whereby diverse subsets of variables within the MIB are made accessible to different management stations without increasing complexity of the managed element.

Finally, a proxy may act as a mid-level manager, supporting aggregated managed objects where the value of one managed object instance depends upon the values of multiple other (remote) items of management information. This can significantly reduce the bandwidth requirements of large-scale management activities.

The architecture of the SNMP network management system is depicted in Figure 2.1. Note that SNMP can also be used between a proxy and its proxied device. For example, an SNMPv2 management station may communicate with an SNMPv1 device via a proxy that uses SNMPv1 protocol.

The Internet-standard network management framework consists of three core technologies as described in Table 2.1.

2.3 Structure of Management Information (SMI)

SMI defines the framework for the definition of SNMP MIBs. SMI describes the common structures and an identification scheme for the definition of management information, particularly an object information model for network management along with a set of generic types used to describe management information. Formal descriptions of the structure are given using a subset of ASN.1 [11, 12].

SNMPv1 SMI (SMIv1) is currently a full Internet standard that is defined in the following RFCs.

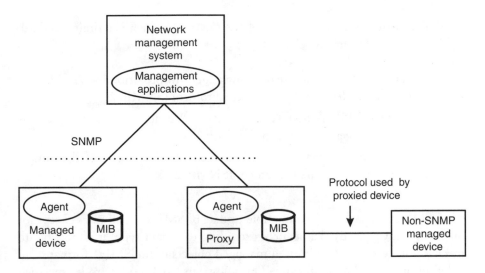

Figure 2.1 SNMP network management system.

Table 2.1
Core Technologies for SNMP

Name	Description
SMI	Describes how managed objects are organized and represented in MIBs.
SNMP	Details the operations of the protocol that is used to exchange management information between SNMP management entities.
MIB	Structured collection of managed objects that represent resources in network elements. By defining MIB modules for new applications and networking techniques, SNMP can be easily extended to support a large variety of devices. In Parts Two and Three, this book will discuss the use of various MIB modules for the management of ATM networks.

- RFC 1155 [1], Structure and Identification of Management Information for TCP/IP-Based Internets;
- RFC 1212 [2], Concise MIB Definitions, which defines a format for producing MIB modules;
- RFC 1215 [13], A Convention for Defining Traps for Use With the SNMP.

The current version is the SNMPv2 (SMIv2), which is a draft standard. Again it consists of three RFCs:

- RFC 1902 [4], Structure of Management Information for Version 2 of the Simple Network Management Protocol (SNMPv2);
- RFC 1903 [5], Textual Conventions for Version 2 of the Simple Network Management Protocol (SNMPv2);
- RFC 1904[6], Conformance Statements for Version 2 of the Simple Network Management Protocol (SNMPv2).

SMIv2 is a backward-compatible update of SMIv1, in all cases except for only one data type. By deliberately excluding that data type, it is possible to mechanically create a definition of managed objects in the SMIv1 format from a definition in the SMIv2 format. This allows an MIB module to be specified in a manner that is both compliant to the SMIv2 and semantically identical to the peer SMIv1 definitions. For example, RFC 1695, which is used extensively in ATM management, is such an MIB.

It should be noted that managed objects are abstractions of resources on systems that exist independently of their need to be managed. The actual realization of these resources is a matter of implementation and is, therefore, out of the scope of any MIB definition.

SMI contains three parts, each defined by an ASN.1 macro, as follows:

- Information module definitions;
- Managed object definitions;
- Notification definitions.

Let us start with the mechanism for identifying managed objects across a network. Note that the following descriptions are based on SMIv2.

2.3.1 Object Definitions

To represent a particular resource uniquely across a network, the object or objects used to identify the resource must be the same at each node that possesses such a resource. This is achieved by using the object identifier (OID) data type defined in ASN.1 [11, 12]. OIDs are organized hierarchically.

2.3.1.1 Object Identifier

Every object is unambiguously identified in SNMP with the assignment of an OID. An OID is an administratively assigned name. However, it can be used for purposes other than naming managed object types; for example, each international standard has an OID assigned to it. In such a case, the OID is just a place holder that helps to organize the OIDs developed further.

An OID consists of a sequence of numbers that identifies the source of the object as well as the object itself. Since this sequence of numbers is variable in length, there is a length field in addition to the sequence of numbers. OIDs are organized in the tree-like structure depicted in Figure 2.2. The sequence of numbers identifies the various branches of the subtree from which a given object is derived. One of the OID root nodes is the ISO; its value is 1. Each branch below the root further identifies the source of the given object. All SNMP objects are members of the subtree identified by iso.org.dod.internet or 1.3.6.1. Note that this is only conventional—there is no reason to limit SNMP MIBs to this part of the tree. Each additional component in this dotted notation further defines the exact location of an object. The numbers for each subtree are assigned by the IETF to ensure that all branches are unique. In fact, all the OIDs shown in Figure 2.2 are place holders; none of them identifies a managed object.

SMI specifies the use of the node internet by defining the following subtrees:

- mgmt { internet 2 }: Identifies objects that are defined in IAB-approved documents. The MIB (1) subtree of this node contains OIDs for all Internet-standard MIBs, including full standard, draft standard, and proposed standard MIBs.

- experimental { internet 3 }: Identifies objects being designed by working groups of the IETF. If an information module produced by a working group becomes a *standard* information module, then at the very beginning of its entry onto the Internet standards track, the objects are moved under the mgmt(2) subtree.

- private { internet 4 }: Identifies objects defined unilaterally. One child of this node, the enterprises(1) subtree, is allocated for enterprises to define their proprietary MIBs. This is a way by which Internet-standard MIBs can be extended. Part Three of this book explains how private MIBs can be used together with standard MIBs to manage an ATM network.

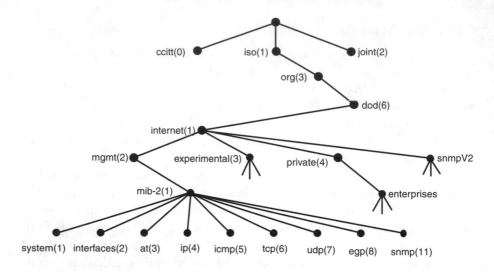

Figure 2.2 OID tree for SNMP.

The division of the internet node into subtrees makes it possible for a smooth evolution of MIBs. A new MIB is usually defined in the experimental or the private subtree to allow vendors and other implementors to experiment with new objects. Consequently, a lot of useful practical experience can be gained, and improper designs can be found and corrected before the MIB is accepted as an Internet standard.

Because OIDs are used so often in SNMP framework, SMIv2 defines a macro, the OBJECT-IDENTITY, to provide information about an OID assignment. All administrative OID assignments that define a type identification value should be defined via the OBJECT-IDENTITY macro.

Definition

```
OBJECT-IDENTITY MACRO ::=
BEGIN
    TYPE NOTATION ::=
                    STATUS    Status
                    DESCRIPTION   Text
                    ReferPart
```

```
VALUE NOTATION ::=
                value(VALUE OBJECT IDENTIFIER)


    Status ::=
                current
        |   deprecated
        |   obsolete


    ReferPart ::=
                REFERENCE   Text
        | empty


    Text ::=    "" string    ""
END
```

Example

```
atmNoTrafficDescriptor   OBJECT-IDENTITY
        STATUS      current
        DESCRIPTION
            This identifies the no ATM traffic
            descriptor type. Parameters 1, 2, 3, 4, and
            5 are not used. This traffic descriptor type
            can be used for best effort traffic.
        ::= { atmTrafficDescriptorTypes 1}
```

2.3.1.2 Object Syntax

Each type of object, termed an object type, has a name and syntax and an encoding. The syntax for an object type defines the abstract data structure corresponding to that object type. ASN.1 constructs are used to define this structure. To avoid complexity, SMI purposely restricts the ASN.1 constructs that may be used. The object syntax is defined by the following ASN.1 definition, and the syntax types that are allowed for use in defining SNMP MIBs are listed in Table 2.2.

Table 2.2
Object Syntax Types

Object Type	SMIv2 RFC 1902	SMIv1 RFC 1155	Description
ASN.1 Built-in Types			
INTEGER (Integer32)	√	√	-2^31 … 2^31-1, may be subtyped to be more constrained. The INTE-GER type may also be used to represent integer-valued information as named-number enumerations.
OCTET STRING	√	√	A sequence of zero or more octets
OBJECT IDENTIFIER	√	√	An ordered list of non-negative numbers assigned administratively.
NULL		√	Commonly used where several alternatives are possible but none of them applies.
Application Wide Types			
IpAddress	√	√	32-bit Internet address
Counter32 (Counter)	√	√	0 … 2^32-1, non-negative integer that monotonically increases until it reaches a maximum value of 2^32-1, when it wraps around and starts increasing again from zero.
Counter64	√		Same as Counter32, but with extended range, i.e.,0 … 2^64-1. Counter64 may be used only if the information being modeled would wrap in less than one hour if the Counter32 type was used instead.
Gauge32 (Gauge)	√	√	A non-negative integer that may increase or decrease but shall never exceed a maximum value, i.e., 2^32-1
TimeTicks	√	√	A non-negative integer that represents the time, modulo 2^32, in hundredths of a second between two epochs.

Object Type	SMIv2	SMIv1	Description
	RFC 1902	RFC 1155	
Application Wide Types			
Opaque	√		Supports the capability to pass arbitrary ASN.1 syntax. Opaque once seemed like a good idea but was quickly dreaded by implementors. This type is provided solely for backward-compatibility and shall not be used for newly defined object types.
Unsigned32	√		Integer valued between 0 and 2^32-1 inclusive.
Network address		√	An address from one of possibly several protocol families. Currently, only one protocol family, the Internet family, is defined. This type is obsolete in current SMIv2.

Definition

```
ObjectSyntax ::=
    CHOICE {
        simple
            SimpleSyntax,


            note that SEQUENCEs for conceptual tables and
            rows are not mentioned here...


        application-wide
            ApplicationSyntax
    }
```

Simple types are built-in types defined by ASN.1. Application-wide types are those defined by standards other than ASN.1. SMI defines a number of application-wide types for SNMP.

Note that a single value of a counter has no information content. The normal SNMP manager application would retrieve the current value of the variable of interest, "remember" it, and then compute delta values using the base value and values obtained from subsequent queries.

2.3.1.3 Definition of Objects

Each managed object type is defined using the OBJECT-TYPE macro, which is explained in Section 2.3.1.3.1.

2.3.1.3.1 OBJECT-TYPE Macro

```
OBJECT-TYPE MACRO ::=
BEGIN
   TYPE NOTATION ::=
                 SYNTAX   Syntax
              UnitsPart
              MAX-ACCESS  Access
              STATUS  Status
              DESCRIPTION  Text
           ReferPart
           IndexPart
           DefValPart


   VALUE NOTATION ::=
                value(VALUE ObjectName)


   Syntax ::=
              type(ObjectSyntax)
           |  BITS   {  Kibbles  }
   Kibbles ::=
           Kibble
           | Kibbles  ,  Kibble
   Kibble ::=
           identifier  (  nonNegativeNumber  )
   UnitsPart ::=
              UNITS   Text
           | empty
   Access ::=
           not-accessible
           | accessible-for-notify
           | read-only
           | read-write
```

```
                |  read-create
Status ::=
                current
                |  deprecated
                |  obsolete
ReferPart ::=
                REFERENCE   Text
                |  empty
IndexPart ::=
                INDEX        {  IndexTypes  }
                |  AUGMENTS   {  Entry       }
                |  empty
IndexTypes ::=
                IndexType
                |  IndexTypes  ,  IndexType
IndexType ::=
                IMPLIED   Index
                |  Index
Index ::=
                use the SYNTAX value of the
                correspondent OBJECT-TYPE invocation
                value(Indexobject ObjectName)
Entry ::=
                use the INDEX value of the
                correspondent OBJECT-TYPE invocation
                value(Entryobject ObjectName)
DefValPart ::=
                DEFVAL   {  value(Defval Syntax)  }
                |  empty
         uses the NVT ASCII character set
Text ::=    "" string    ""
END
```

Syntax. This mandatory clause defines the abstract data structure corresponding to an object. The data structure must be one of the following: a base type,
the BITS construct, or a textual convention. The base types are those defined in
the ObjectSyntax described earlier. A textual convention is an item used to
specify additional semantics to an existing syntax type, as we will see in
Section 2.3.4. The BITS construct represents an enumeration of named bits.

UNITS. The UNITS is an optional clause that contains a textual definition of the units associated with that object, for example, seconds. The information provided by this clause makes a management station's user-interface more user-friendly.

MAX-ACCESS. The MAX-ACCESS clause defines the maximal level of access for the object. This part must be present, as it defines whether it makes "protocol sense" to read, write, and/or create an instance of the object, or to include its value in a notification. The values for MAX-ACCESS are ordered, from least to greatest, as follows:

- not-accessible: May not be directly read, written, or created;
- accessible-for-notify: Accessible only via a notification (e.g., snmpTrapOID);
- read-only: May be read, but not written or created;
- read-write: May be read or written, but not created;
- read-create: May be read, written, and created. This value is used for writeable objects in a conceptual row for which new instances can be created via network management.

STATUS. The STATUS clause, which must be present, indicates whether this definition is current or historic. The values *current* and *obsolete* are self-explanatory. The *deprecated* value indicates that the definition is obsolete but that an implementor may wish to support that object to foster interoperability with older implementations.

DESCRIPTION. This mandatory clause contains a textual definition of the object that provides all semantic definitions necessary for implementation and should embody any information that would otherwise be communicated in any ASN.1 commentary annotations associated with the object.

REFERENCE. This optional clause contains a textual cross-reference to an object defined in some other information module.

DEFVAL. This optional clause defines an acceptable default value that may be used at the discretion of an SNMPv2 entity acting in an agent role when an object instance is created.

OBJECT-TYPE. The value of an invocation of the OBJECT-TYPE macro is the name of the object, which is an OBJECT IDENTIFIER, an administratively assigned name.

INDEX and AUGMENTS. The mapping of these two clauses will be introduced in Section 2.3.1.3.3.

2.3.1.3.2 Scalar Objects

The Object-Type macro can be used to define scalar objects, also known as leaf objects, and conceptual tables. An example of defining a scalar object is given as follows:

```
atmVpCrossConnectIndexNext    OBJECT-TYPE
        SYNTAX  INTEGER (0..2147483647)
        MAX-ACCESS  read-only
        STATUS  current
        DESCRIPTION
        This object contains an appropriate value to
        be used for atmVpCrossConnectIndex when
        creating entries in the
        atmVpCrossConnectTable. The value 0 indicates
        that no unassigned entries are available. To
        obtain the atmVpCrossConnectIndex value for a
        new entry, the manager issues a management
        protocol retrieval operation to obtain the
        current value of this object. After each
        retrieval, the agent should modify the value
        to the next unassigned index.
        ::= { atmMIBObjects 8 }
```

2.3.1.3.3 Conceptual Tables

In SNMP, management operations apply exclusively to scalar objects. However, it is sometimes convenient for developers of management applications to impose an imaginary, tabular structure on an ordered collection of objects within the MIB. Each such conceptual table contains zero or more rows, and each row may contain one or more scalar objects, termed *columnar objects*. To construct a conceptual table, two ASN.1 structured types are used:

- *Sequence:* A structured type defined by referencing a fixed, ordered list of types (some of which may be declared to be optional); each value of the new type is an ordered list of values, one from each component type.

- *Sequence of:* A structured type defined by referencing a single existing type; each value in the new type is an ordered list of zero, one or more values of the existing type.

This conceptual table is formalized by using the OBJECT-TYPE macro to define both an object that corresponds to a table and an object that corresponds to a row in that table.

Conceptual row. A conceptual row has SYNTAX of the form:

```
<Entry Type>
where <Entry Type> is a SEQUENCE type defined as follows:
<Entry Type> ::= SEQUENCE { <type1>, ... , <typeN> }
```

where there is one <type> for each subordinate object, and each <type> is a leaf object, which was described earlier.

An example of a conceptual row follows.

```
atmInterfaceDs3PlcpEntry    OBJECT-TYPE
          SYNTAX      AtmInterfaceDs3PlcpEntry
          MAX-ACCESS  not-accessible
          STATUS      current
          DESCRIPTION
            This list contains DS3 PLCP parameters and
            states variables at the ATM interface.
          INDEX    {ifIndex }
          ::= { atmInterfaceDs3PlcpTable 1}


AtmInterfaceDs3PlcpEntry    ::= SEQUENCE  {
          atmInterfaceDs3PlcpSEFSs          Counter32,
          atmInterfaceDs3PlcpAlarmState     INTEGER,
          atmInterfaceDs3PlcpUASs           Counter32
          }
```

Conceptual table. A conceptual table has SYNTAX of the form:

```
SEQUENCE OF <EntryType>
```

where `<Entry Type>` refers to the SEQUENCE type of its subordinate conceptual row.

An example of a conceptual table follows.

```
atmInterfaceDs3PlcpTable    OBJECT-TYPE
    SYNTAX        SEQUENCE OF AtmInterfaceDs3PlcpEntry
    MAX-ACCESS    not-accessible
    STATUS        current
    DESCRIPTION
     This table contains ATM interface DS3 PLCP
    parameters and state variables, one entry per
    ATM interface port.
    ::= { atmMIBObjects 3}
```

SNMP states that management operations apply exclusively to scalar objects. Consequently, the MAX-ACCESS clause for conceptual tables and rows is "not-accessible," because they are not directly accessible by management protocol. Conceptual row creation/deletion will be discussed in Section 2.5.1.

A row in a conceptual table is indexed by the INDEX or AUGMENTS clause. The AUGMENTS clause is an alternative to the INDEX clause. The introduction of the AUGMENTS clause allows the related management information to be defined in several MIB modules. For instance, a well-established standard MIB may contain a table for which additional columns need to be defined in a new MIB, either a potential future standard or an enterprise-specific MIB. In particular, the AUGMENTS clause replaces the INDEX clause to indicate that the conceptual rows in the new table are identified in the same way as the conceptual rows in the original table. Note that the AUGMENTS clause should only be used if there is a one-to-one correspondence between the conceptual rows of this table and an existing table that is indexed by an INDEX clause.

Table 2.3 is an example of object instances of the DS3 PLCP table defined in the AToM MIB, which will be introduced in Chapter 6. Table 2.3 contains three columnar objects: atmInterfaceDs3PlcpSEFSs, atmInterfaceDs3PlcpAlarmState, and atmInterfaceDs3PlcpUASs, containing DS3 PLCP configuration and state parameters of those ATM interfaces that use DS3 PLCP for carrying ATM cells over DS3. Please refer to Chapter 6 for a detailed description of these objects. Table 2.3 is indexed by the interface index (ifIndex); the ATM device referenced by Table 2.3 has three ATM interfaces with ifIndex equal to 1, 2, and 3.

Table 2.3
A Snapshot of the DS3 PLCP Table

Index	Columnar Objects		
ifIndex	atmInterface-Ds3PlcpSEFSs	atmInterface-Ds3PlcpAlarmState	atmInterface-Ds3PlcpUASs
1	0	noAlarm(1)	0
2	27	noAlarm(1)	5
3	235	noAlarm(1)	40

2.3.1.4 Instance Identification

The OBJECT-TYPE macro defines types of managed objects. However, it is the instances of objects that are manipulated by a management protocol. SMI also defines the roles for identifying object instances.

Basically, an instance of an object is identified by the OID of its type and a suffix. The resulting identifier is another OID. By naming instances using OIDs, a lexicographic ordering is enforced over all object instances. This feature is used in automatically detecting the structure of an MIB as will be introduced in Section 2.4.4.2.

For leaf objects that are not columnar objects (i.e., not contained within a conceptual table), instances of the object are identified by appending a subidentifier of zero to the name of that object.

For example, the instance identifier for sysUpTime is

sysUpTime.0 or 1.3.6.1.2.1.1.3.0

For columnar objects, there are potentially multiple instances, each corresponding to a cell in a given row in the table. Conceptual rows are identified by the INDEX or AUGMENTS clause, as are the instances in the row. Consequently, the suffix for a columnar object consists of the objects that are specified in the INDEX clause. The detailed rules are listed in Table 2.4.

Table 2.5 shows the OIDs for instances of the objects listed in Table 2.3. Moreover, Figure 2.3 depicts the OID tree for the instances and their lexicographic ordering.

Table 2.4
Instance Encoding Rules for Columnar Objects

Syntax Type	Suffix Encoding
Integer	A single subidentifier taking the integer value (this works only for non-negative integers)
Fixed-length string	*n* sub-identifiers, where *n* is the length of the string (each octet of the string is encoded in a separate subidentifier)
Variable-length string	*n* + 1 subidentifiers, where *n* is the length of the string (the first subidentifier is *n* itself; following this, each octet of the string is encoded in a separate subidentifier)
IpAddress	Four subidentifiers, in the well-known a.b.c.d notation
OID	*n* + 1 subidentifiers, where *n* is the number of subidentifiers in the value (the first subidentifier is *n* itself, following this, each subidentifier in the value is copied)

2.3.2 Module Definitions

The MODULE-IDENTITY macro is used to provide contact and revision history for each information module.

All information modules start with exactly one invocation of the MODULE-IDENTITY macro, which provides contact information as well as revision history to distinguish between versions of the same information module. This invocation must appear immediately after any IMPORTs statements.

Definition

```
MODULE-IDENTITY MACRO ::=
BEGIN
    TYPE NOTATION ::=
            LAST-UPDATED   value(Update UTCTime)
            ORGANIZATION   Text
            CONTACT-INFO   Text
            DESCRIPTION    Text
            RevisionPart
    VALUE NOTATION ::=
            value(VALUE OBJECT IDENTIFIER)
```

Table 2.5
Example of Instance OIDs for the DS3 PLCP Table

MIB Object	Index	Object Instance	
	ifIndex	OID	Value
atmInterfaceDs3PlcpSEFSs	1	atmInterfaceDs3PlcpSEFSs.1	0
	2	atmInterfaceDs3PlcpSEFSs.2	27
	3	atmInterfaceDs3PlcpSEFSs.3	235
atmInterfaceDs3PlcpAlarmState	1	atmInterfaceDs3PlcpAlarmState.1	noAlarm(1)
	2	atmInterfaceDs3PlcpAlarmState.2	noAlarm(1)
	3	atmInterfaceDs3PlcpAlarmState.3	noAlarm(1)
atmInterfaceDs3PlcpUASs	1	atmInterfaceDs3PlcpUASs.1	0
	2	atmInterfaceDs3PlcpUASs.2	5
	3	atmInterfaceDs3PlcpUASs.3	40

```
RevisionPart ::=
        Revisions
    | empty

Revisions ::=
        Revision
    | Revisions Revision
Revision ::=
        REVISION  value(Update UTCTime)
        DESCRIPTION  Text


    uses the NVT ASCII character set
    Text ::=   "" string   ""
END
```

Example

```
atmMIB MODULE-IDENTITY
        LAST-UPDATED  9406072245Z
```

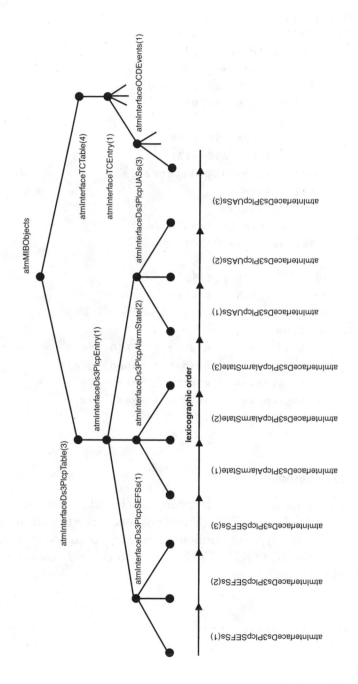

Figure 2.3 Instance OID and lexicographic ordering.

```
ORGANIZATION   IETF AToM MIB Working Group
CONTACT-INFO
      Masuma Ahmed
   Postal:  Bellcore
      331 Newman Springs Road
      Red Bank, NJ 07701
      US
   Tel:       +1 908 758 2515
   Fax:       +1 908 758 4131
   E-mail:  mxa@mail.bellcore.com
      Kaj Tesink
   Postal:  Bellcore
      331 Newman Springs Road
      Red Bank, NJ 07701
      US
   Tel:       +1 908 758 5254
   Fax:       +1 908 758 4196
   E-mail:  kaj@cc.bellcore.com
DESCRIPTION
   "This is the MIB Module for ATM and
   AAL5-related objects for managing ATM
   interfaces, ATM virtual links, ATM
   cross-connects, AAL5 entities, and
   AAL5 connections."
         ::= { mib-2 37 }
```

2.3.3 Notification Definitions

Notification definitions are used when describing unsolicited transmissions of management information. In SNMPv1, there is only one type of notification, i.e., traps. SNMPv2 has two types of notifications, namely, traps and informs. An ASN.1 macro, NOTIFICATION-TYPE, is used to concisely convey the syntax and semantics of a notification.

Definition

```
NOTIFICATION-TYPE MACRO ::=
BEGIN
   TYPE NOTATION ::=
         ObjectsPart
```

```
            STATUS    Status
            DESCRIPTION   Text
            ReferPart


    VALUE NOTATION ::=
            value(VALUE NotificationName)


    ObjectsPart ::=
            OBJECTS   { Objects  }
        | empty
    Objects ::=
            Object
        | Objects  ,  Object
    Object ::=
            value(Name ObjectName)


    Status ::=
            current
        | deprecated
        | obsolete


    ReferPart ::=
            REFERENCE   Text
        | empty


        uses the NVT ASCII character set
        Text ::=   "" string   ""
    END
```

Example

```
coldStart NOTIFICATION-TYPE
    STATUS  current
    DESCRIPTION
        A coldStart trap signifies that the SNMPv2 entity,
        acting in an agent role, is reinitializing itself and
```

```
    that its configuration may have been altered.
::= { snmpTraps 1 }
```

2.3.4 Textual Conventions

When designing an MIB module, it is often useful to define new types similar to those defined in the SMI. In comparison to a type defined in the SMI, each of these new types has a different name, a similar syntax, but a more precise semantic. These newly defined types are termed *textual conventions* and are used for the convenience of humans reading the MIB module. RFC 1903, "Textual Conventions for Version 2 of the Simple Network Management Protocol (SNMPv2) Framework," defines the initial set of textual conventions available to all SNMPv2 MIB modules. Enterprises may define their own textual conventions for their private MIBs.

Objects defined using a textual convention are always encoded by means of the rules that define their primitive type. However, textual conventions often have special semantics associated with them. As such, an ASN.1 macro, TEXTUAL-CONVENTION, is used to concisely convey the syntax and semantics of a textual convention.

Definitions

```
    SNMPv2-TC DEFINITIONS ::= BEGIN


    IMPORTS
        ObjectSyntax, TimeTicks
            FROM SNMPv2-SMI;
        definition of textual conventions


    TEXTUAL-CONVENTION MACRO ::=
    BEGIN
        TYPE NOTATION ::=
                DisplayPart
                STATUS    Status
                DESCRIPTION    Text
                ReferPart
                SYNTAX    Syntax
```

```
VALUE NOTATION ::=
        value(VALUE Syntax)


DisplayPart ::=
        DISPLAY-HINT  Text
    | empty


Status ::=
        current
    | deprecated
    | obsolete


ReferPart ::=
        REFERENCE  Text
    | empty


  uses the NVT ASCII character set
Text ::=    "" string    ""


Syntax ::=
        type(ObjectSyntax)
    | BITS   {  Kibbles  }
Kibbles ::=
        Kibble
    | Kibbles  ,  Kibble
Kibble ::=
        identifier (  nonNegativeNumber  )
END
```

Example

```
AtmTrafficDescrParamIndex ::= TEXTUAL-CONVENTION
        STATUS      current
        DESCRIPTION
          The value of this object identifies the row
          in the atmTrafficDescrParamTable.
        SYNTAX      Integer32
```

Table 2.6
List of Textual Conventions Defined in RFC 1902

Name	Definition
DisplayString	An NVT ASCII string up to 255 characters
PhysAddress	Represents media- or physical-level addresses
MacAddress	48-bit IEEE 802 MAC address
TruthValue	Boolean value
TestAndIncr	Provides atomic operations. It can be used not only to ensure that multiple set operations are executed at most once and in the desired order, but also to provide advisory locking between multiple management applications
AutonomousType	An independently extensible type identification value
InstancePointer	A pointer to either a specific instance of an MIB object or a conceptual row of an MIB table in the managed device
VariablePointer	A pointer to a specific object instance
RowPointer	A pointer to a conceptual row
RowStatus	Used to manage the creation and deletion of conceptual rows
TimeStamp	The value of the sysUpTime object at which a specific occurrence happened
TimeInterval	A period of time, measured in units of 0.01 seconds
DateAndTime	A date-time specification
StorageType	Describes the memory used for a conceptual row
TDomain	Denotes a kind of transport service
TAddress	Denotes a transport service address

2.3.5 Conformance

Due to various reasons, such as limited memory, processing power, or cost, an agent may not fully support all the objects defined by MIBs that it implements. This may, in turn, result in an implementation that is not interoperable with management applications. To avoid this problem, it is very useful to define a set of acceptable lower bounds of implementation, along with the actual level of implementation achieved. RFC 1904, "Conformance Statements for Version 2 of the Simple Network Management Protocol (SNMPv2)," defines the notation used for these purposes.

In particular, RFC 1904 specifies the definitions for the following macros:

- OBJECT-GROUP macro;
- NOTIFICATION-GROUP macro;
- MODULE-COMPLIANCE macro;
- AGENT-CAPABILITIES macro.

2.3.5.1 OBJECT-GROUP Macro

OBJECT-GROUP macro is used to specify a collection of related managed objects for conformance purposes. This macro defines groups in a more formal manner and allows the contained objects to be specified explicitly, no matter where they are positioned within the naming structure of an MIB. It is required that every object defined in an information module with a MAX-ACCESS clause other than "not-accessible" be contained in at least one object group.

Definitions

```
OBJECT-GROUP MACRO ::=
BEGIN
  TYPE NOTATION ::=
          ObjectsPart
           STATUS   Status
           DESCRIPTION   Text
           ReferPart
  VALUE NOTATION ::=
           value(VALUE OBJECT IDENTIFIER)
  ObjectsPart ::=
           OBJECTS  {  Objects  }
  Objects ::=
           Object
         | Objects  ,  Object
  Object ::=


  value(Name ObjectName)
  Status ::=
           current
         | deprecated
         | obsolete
  ReferPart ::=
           REFERENCE   Text
         | empty
```

```
        uses the NVT ASCII character set
    Text ::=    "" string    ""
END
```

Example

```
atmVpcTerminationGroup     OBJECT-GROUP
    OBJECTS {atmVplOperStatus, atmVplAdminStatus,
        atmVplLastChange,
        atmVplReceiveTrafficDescrIndex,
        atmVplTransmitTrafficDescrIndex,
        atmVplRowStatus }
    STATUS      current
    DESCRIPTION
        A collection of objects providing
        information about a VPL at an ATM interface
        which terminates a VPC
        (i.e., one which is NOT cross-connected
        to other VPLs)."
    ::= { atmMIBGroups 5 }
```

2.3.5.2 NOTIFICATION-GROUP Macro

For conformance purposes, it is also useful to define a collection of notifications. The NOTIFICATION-GROUP macro serves this purpose. This
macro allows the MODULE-COMPLIANCE macro to refer to groups
of notifications and the AGENT-CAPABILITIES macro to define variations on implementing notifications. The definition and usage of the
NOTIFICATION-GROUP macro corresponds exactly to the OBJECT-
GROUP macro.

2.3.5.3 MODULE-COMPLIANCE Macro

The MODULE-COMPLIANCE macro is used when describing requirements
for agents with respect to managed objects. As mentioned in 2.3.1.3, object
definitions in an MIB module define the maximum level of implementation
that makes protocol sense. In contrast, the module conformance macro defines
the minimum requirements for conformance, specified in terms of objects
within groups, where different groups may come from different MIB modules.
A requirement on all "standard" MIB modules is that a corresponding
MODULE-COMPLIANCE specification also be defined, either in the same
information module or in a companion information module.

The MODULE-COMPLIANCE macro lists all the objects and groups used in the MIB. The access to these objects can be specified by the MIN-ACCESS clause, which defines the minimum access level. In addition, a minimum set of managed object groups is specified as "mandatory" by this macro. Any SNMPv2 compliant implementation must implement such groups.

2.3.5.4 AGENT-CAPABILITIES Macro

The AGENT-CAPABILITIES macro is used to convey a set of capabilities present in an SNMPv2 entity acting in an agent role. When an MIB module is written, it is divided into units of conformance-termed groups. If an SNMPv2 entity acting in an agent role claims to implement a group, then it must implement each and every object within that group. Of course, for whatever reason, an SNMPv2 entity might implement only a subset of the groups within an MIB module. In addition, the definition of some MIB objects may leave some aspects of the definition to the discretion of an implementor.

Practical experience has demonstrated a need for concisely describing the capabilities of an agent with respect to one or more MIB modules. The AGENT-CAPABILITIES macro allows an agent implementor to describe the precise level of support that an agent claims in regards to an MIB group. This information is useful to optimize applications that manage such an agent.

Note that the AGENT-CAPABILITIES macro specifies refinements or variations with respect to OBJECT-TYPE and NOTIFICATION-TYPE in MIB modules, *not* with respect to MODULE-COMPLIANCE macros in compliance statements.

2.3.6 SMI Summary

SMI describes how to use a subset of ASN.1 to define management information. Typically, there are three kinds of information modules as follows:

- MIB modules, which contain definitions of inter-related managed obects and make use of the OBJECT-TYPE and NOTIFICATION-TYPE macros;
- Compliance statements for MIB modules, which make use of the MODULE-COMPLIANCE and OBJECT-GROUP macros;
- Capability statements for agent implementations, which make use of the AGENT-CAPABILITIES macros.

It is not necessary for an information module to contain all three modules. A *standard* MIB normally includes only definitions of managed objects and a compliance statement. Similarly, an *enterprise-specific* information module might include definitions of managed objects and, in a product-specific module, provide a capability statement.

2.4   SNMP Protocol

SNMP protocol is used to convey management information between agents and management stations, as well as between different management stations.

2.4.1   Protocol Interactions

Three types of access to management information are provided by the SNMP protocol. They are described as follows.

1. Manager-agent request-response interaction: an SNMP entity acting in a manager role sends a request to an SNMP entity acting in an agent role, and the latter SNMP entity responds to the request. This type is used to retrieve or modify management information associated with the managed device.

2. Manager-manager request-response interaction: an SNMP entity acting in a manager role sends a request to an SNMP entity also acting in a manager role, and the latter SNMPv2 entity then responds to the request. This type is used to notify an SNMP entity acting in a manager role of management information associated with another SNMP entity also acting in a manager role.

3. Unconfirmed interaction: an SNMP entity acting in an agent role sends an unsolicited message, termed a trap, to an SNMP entity acting in a manager role, and no response is returned. This type is used to notify an SNMP entity acting in a manager role of an exceptional situation that has resulted in changes to management information associated with the managed device.

Table 2.7 summarizes all operations supported by the SNMPv2 protocol, which will be introduced in turn.

The native protocol to carry protocol information for SNMP is the connectionless *user datagram protocol* (UDP), although mapping into other protocols is possible. By convention, SNMP agents listen on UDP port 161, and notifications are sent to UDP port 162. Every message is entirely and independently represented by a single UDP datagram. The argument for the choice

Table 2.7

Summary of SNMP Protocol Operations

Involved Entities	Interaction Type	Request	Response	Trap
Manager-agent	Request-response	get-request	response	
		get-next-request		
		get-bulk-request		
		set-request		
	Unconfirmed			snmp trap
Manager-manager		inform-request	response	

of connectionless UDP is that SNMP must continue to operate, if at all possible, when the network is operating at its worst. The major purpose of a network management protocol is to fix the network when it is operating poorly. In such cases, connection-oriented protocols (i.e., TCP) may not work properly. With the use of UDP, the complexity of the management agent and the resources required can be minimized.

SNMP must be able to receive messages of at least 484 octets in size. Larger maximum values are permitted, via bilateral agreement. The relatively small message size was a goal of the design but for some reasonable sets of network management functions, it imposes a limitation.

2.4.2 PDU Formats

SNMP defines eight different data types for its *packet data units* (PDUs) by using only two different formats, i.e., PDU and BulkPDU as listed in Table 2.8. The format of PDU is shown in Figure 2.4; its fields are explained in Tables 2.9 and 2.10. The BulkPDU will be introduced in Section 2.4.4.3.

2.4.3 MIB View

For security reasons, it is often valuable to be able to restrict the access rights of some management applications to only a subset of the management information in the management domain. To provide this capability, the concept of MIB view is used to restrict the management information a management station can access.

Table 2.8
PDU Overview

PDU Type	Format
GetRequest-PDU	PDU
GetNextRequest-PDU	
Response-PDU	
SetRequest-PDU	
InformRequest-PDU	
SNMPv2-Trap-PDU	
Report-PDU*	
GetBulkRequest-PDU	BulkPDU

* For potential future use to report errors between SNMPv2
entities rather than between SNMPv2 managers and agents.

Table 2.9
PDU Format

Field Name	Usage
request-id	An integer value used by a management application to distinguish among outstanding requests.
error-status	A non-zero value of the error-status field in a Response-PDU is used to indicate that an exception occurred to prevent the processing of the request. In Request-PDUs, this field should be set to zero.
error-index	A non-zero value of the Response-PDU's error-index field provides additional information by identifying which variable binding in the list caused the exception. Details of error codes are listed in Table 2.10. In Request-PDUs, this field should be set to zero.
variable-bindings	A variable binding is identified by its index value. The first variable binding in a variable-binding list is index one; the second is index two, etc.

| request-id | error status | error-index | variable-bindings |

Figure 2.4 PDU format.

Table 2.10
Error codes for error-status field

Name	Code
noError	0
tooBig	1
noSuchName	2
badValue	3
readOnly	4
genErr	5
noAccess	6
wrongType	7
wrongLength	8
wrongEncoding	9
wrongValue	10
noCreation	11
inconsistentValue	12
resourceUnavailable	13
commitFailed	14
undoFailed	15
authorizationError	16
notWritable	17
inconsistentName	18

Since managed object types and their instances are identified via the tree-like naming structure, it is convenient to define an MIB view as the combination of a set of *view subtrees,* where each view subtree is a subtree within the managed object naming tree. Thus, a simple MIB view (e.g., all managed objects within the Internet network management framework) can be defined as a single view subtree, while more complicated MIB views (e.g., all information relevant to a particular network interface) can be represented by the union of multiple view subtrees.

In addition to restricting access rights by identifying subsets of management information, it is also valuable to restrict the requests allowed on the management information within a particular device. For example, one management application might be prohibited from write-access to a particular device, while another might be allowed to perform any type of operation.

2.4.4 Retrieving Operations

Retrieving operations are initiated by management applications to retrieve management information from agents. There are three such operations defined in SNMP—get-request, get-next-request, and get-bulk-request. See Figure 2.5.

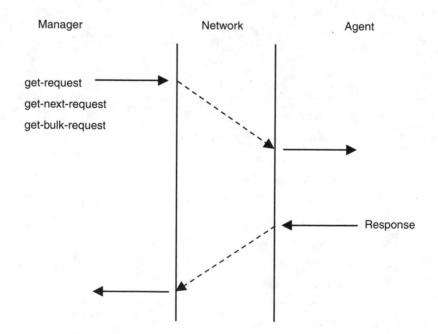

Figure 2.5 Retrieving operations.

2.4.4.1 get-request

A GetRequest-PDU is generated and transmitted at the request of an SNMP application, i.e., a manager. Upon receipt of a GetRequest-PDU, the receiving SNMP entity performs a series of processing to generate a Response-PDU to be sent back to the originator as follows:

Request-id

This field is always the same as the received GetRequest-PDU.

Variable Binding Processing

Upon receipt of a GetRequest-PDU, the receiving SNMP entity, i.e., an agent, processes each variable binding in the variable-binding list to produce a Response-PDU. Each variable binding is processed as follows:

1. If the variable binding's name exactly matches the name of a variable accessible by this request, then the variable binding's value field is set to the *value* of the named variable

2. If the variable binding's name does not have an OID prefix that exactly matches the OID prefix of any (potential) variable accessible by this request, then its value field is set to noSuchObject.

3. Otherwise, the variable binding's value field is set to noSuch-Instance.

Error-Status/Error-Index

1. If no error is encountered, the error-status field should be set to noError and the error-index set to zero.

2. If the size of the generated message is greater than or equal to a local constraint or the maximum message size of the originator, it is transmitted to the originator of the GetRequest-PDU, then an alternate Response-PDU is generated. This alternate Response-PDU is formatted with the value of its error-status field set to tooBig, the value of its error-index field set to zero, and an empty variable-bindings field.

3. If any other error occurs, then the Response-PDU is reformatted with the same variable-bindings fields as the received GetRequest-PDU. The value of its error-status field set to genErr, and the value of its error-index field is set to the index of the failed variable binding.

It should be noted that get-request can only be used when the manager knows exactly the instances of the management information it wants.

2.4.4.2 get-next-request

The syntax of get-next-request is almost the same as that of get-request. The difference is that get-next-request permits the viewing of data without requiring prior knowledge. As discussed in Section 2.3, every instance of managed object is uniquely identified by an OBJECT IDENTIFIER that is located in the lexicographically ordered list of the names of all variables. This makes it possible for the management information to be traversed in lexicographical order.

In contrast to get-request, the processing of get-next-request does not necessarily require that the variable binding's name exactly match the name of a variable accessible by this request. The variable binding processing in get-next-request is as follows:

1.   The variable is located within the lexicographically ordered list of the names of all variables that are accessible by this request, and its name is the first lexicographic successor of the variable binding's name in the incoming GetNextRequest-PDU. The corresponding variable binding's name and value fields in the Response-PDU are set to the name and value of the located variable.

2.   If the requested variable binding's name does not lexicographically precede the name of any variable accessible by this request (i.e., there is no lexicographic successor), then the corresponding variable binding produced in the Response-PDU has its value field set to endOf-MibView, and its name field set to the variable binding's name in the request.

Example of Table Traversal

Let us examine how to retrieve the object instances in Figure 2.3 by making use of get-next-request. The SNMP manager begins by sending a GetNextRequest-PDU containing the indicated OID values as the requested variable names:

```
get-next-request (
    sysUpTime,
    atmInterfaceDs3PlcpSEFSs,
    atmInterfaceDs3PlcpUASs)
```

The agent responds with a Response-PDU:

```
response (
   sysUpTime.0 =    123456" ),
   atmInterfaceDs3PlcpSEFSs.1 = 0,
   atmInterfaceDs3PlcpUASs.1 =  0 )
```

The manager continues with:

```
get-next-request (
   sysUpTime,
   atmInterfaceDs3PlcpSEFSs.1,
   atmInterfaceDs3PlcpUASs.1)
```

The agent responds with a Response-PDU:

```
response (
   sysUpTime.0 =    123461",
   atmInterfaceDs3PlcpSEFSs.2 = 27 ,
   atmInterfaceDs3PlcpUASs.2 =  5  )
```

The manager continues with:

```
get-next-request (
   sysUpTime,
   atmInterfaceDs3PlcpSEFSs.2,
   atmInterfaceDs3PlcpUASs.2)
```

The agent responds with a Response-PDU:

```
response (
   sysUpTime.0 =    123466" ,
   atmInterfaceDs3PlcpSEFSs.3 = 235,
   atmInterfaceDs3PlcpUASs.3 =   40)
```

The manager continues with:

```
get-next-request (
   sysUpTime,
   atmInterfaceDs3PlcpSEFSs.3,
   atmInterfaceDs3PlcpUASs.3)
```

As there are no further entries in the table, the agent responds with the variables that are next in the lexicographical ordering of the accessible object names. For example:

```
response (
   sysUpTime.0 =     123471" ,
   atmInterfaceDs3PlcpAlarmState.1    = 1,
   atmInterfaceOCDEvents.1 =   5 )
```

The manager sees that the last two variable bindings do not correspond to the information it wanted, so it knows that all instances of the table in question have been retrieved.

The get-next is capable of detecting the structure of a whole MIB. Therefore, it is often called *powerful* get-next. However, unless employed with care, the get-next primitive can be extremely resource-intensive in terms of network bandwidth and CPU time.

2.4.4.3   get-bulk-request

The get-bulk-request is probably the most successful change that SNMPv2 has made in its evolution from SNMPv1. This operator greatly improves the performances of bulk data retrieving (i.e., large tables). The basic concept is that, with one PDU, get-bulk performs repeated get-next executions. The format of GetBulkRequest-PDU is shown in Figure 2.6, and various fields are explained in Table 2.11.

Two parameters are included in the request: non-repeaters and max-repetitions as explained in Table 2.11. Consequently, the total number of requested variable bindings communicated by the request is given by $N + (M \times R)$. This design is useful when the values of one or more scalars (e.g., sysUpTime) must be retrieved along with variables from multiple rows of a table, in that only one copy of the sysUpTime is retrieved along with the multiple rows from the table.

Thus, through using get-bulk, a manager can retrieve in one request as many variables (e.g., from the rows of a large table) as will fit in a maximum-sized response without knowing the names (i.e., the instance identifiers) of the particular variables it wants to retrieve. When max-repetition has a value of

request-id	non-repeaters	max-repetitions	variable-bindings

Figure 2.6   Bulk PDU format.

Table 2.11
Fields in Bulk PDU

Field Name	Usage
request-id	An integer value used by a management application to distinguish among outstanding requests.
non-repeaters	The number (N) of variables beginning from first to the Nth variable in the variable-bindings that should be retrieved at most once.
max-repetitions	The maximum number (M) of times that remaining variables (R) in the variable-bindings should be retrieved.
variable bindings	A variable binding is identified by its index value. The first variable binding in a variable-binding list is index one, the second is index two, etc.

one, get-bulk operates identically to get-next, with the exception that get-bulk never returns a tooBig error; if a get-next would return tooBig, get-bulk will return less than a whole repetition.

The following example of the DS3 PLCP table using get-bulk-request demonstrates how awesome this operator is.

```
get-bulk-request [ non-repeaters = 1, max-repetitions = 2
] (
    sysUpTime,
    atmInterfaceDs3PlcpAlarmState,
    atmInterfaceDs3PlcpUASs )
```

The agent responds with a Response-PDU:

```
response (
    sysUpTime.0 =    123456",
    atmInterfaceDs3PlcpSEFSs.1 = 0,
    atmInterfaceDs3PlcpUASs.1 =  0,
    atmInterfaceDs3PlcpSEFSs.2 = 27 ,
    atmInterfaceDs3PlcpUASs.2 =  5 )
```

The manager continues with:

```
get-bulk-request [ non-repeaters = 1, max-repetitions = 2
] (
```

```
sysUpTime,
atmInterfaceDs3PlcpSEFSs.2,
atmInterfaceDs3PlcpUASs.2)
```

The agent responds:

```
response (
    sysUpTime.0 =    123466" ,
    atmInterfaceDs3PlcpSEFSs.3 = 235,
    atmInterfaceDs3PlcpUASs.3 =    40,
    atmInterfaceDs3PlcpAlarmState.1 = 1,
    atmInterfaceOCDEvents.1 =   5 )
```

This response signals the end of the table to the manager.

Although the awesome get-bulk operator is highlighted for its blazing speed, it also reduces the complexity of SNMP applications. The get-bulk operation fills up a packet with variable-bindings until the packet is full. There is no danger of returning a tooBig error with the get-bulk operator. This obviates the need for code in the application that, upon receipt of such a reply, splits up the request packet into smaller chunks. In addition, there is no longer any need to have the management station dynamically discover (by trial and error) the optimal number of variable-bindings to put in the request packet to get the most data in the reply packet. Interestingly, the introduction of the operator requires remarkably few changes to an SNMPv1 agent [14, 15].

2.4.5 set-request

In SNMP, there is only one way to change the value of a specified managed object instance—that is, by issuing a set-request. (See Figure 2.7.) SNMP deliberately excludes imperative commands from the set of explicitly supported management functions to keep its protocol simple, because the number of such commands is, in practice, ever-increasing, and the semantics of such commands are generally arbitrarily complex. With set-request, an imperative command can be realized as the setting of a parameter value that subsequently triggers the desired action. For example, rather than implementing a *reboot command*, this action might be invoked by simply setting a parameter indicating the number of seconds until system reboot.

Upon receipt of a SetRequest-PDU, the agent first estimates the size of a resulting Response-PDU. If the estimated message size is greater than either a local constraint or the maximum message size of the request's source party,

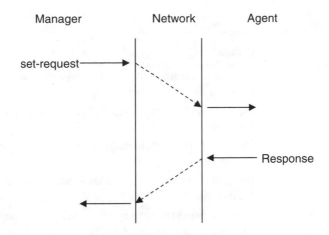

Manager   Network   Agent

set-request ────▶

                                  ──────▶

                          ◀────── Response

◀──────

Figure 2.7 set-request operation.

then an alternate Response-PDU is generated, with the value of its error-status field set to tooBig, the value of its error-index field set to zero, and an empty variable-bindings field.

Otherwise, the agent starts to process the variable bindings in two phases. In the first phase, each variable binding is validated; if all validations are successful, then each variable is altered in the second phase.

Each variable binding is validated in the first phase until all variable bindings are successful, or until one fails. If any failure occurs, the error status field should be set to an error code according to Table 2.12, and the value of its error-index is set to the index of the failed variable binding. At the end of the first phase, if the validation of all variable bindings succeeded, then the value of the Response-PDU's error-status field is set to noError and the value of its error-index field is zero.

Table 2.12
Error Code for set-request Operation

Error Code	Reason
noAccess	The variable is outside the MIB view for the operation
notWritable	The agent is unable to modify/create the instances of the variable binding's object type
wrongType	If the type of the variable binding's value field is inconsistent with the variable binding's object type

Table 2.12 (continued)

Error Code	Reason
wrongLength	Length of the variable binding's value field is inconsistent with the variable binding's object type
wrongEncoding	The encoding of the variable binding's value field is inconsistent with the variable binding's object type
wrongValue	The value of the variable binding's value field could not be assigned to its instance
noCreation	The variable binding's name specifies a variable that does not exist and could never be created
inconsistentName	The variable binding's name specifies a variable that cannot be created under the present circumstances
inconsistentValue	The variable binding's value field specifies a value that is presently inconsistent or otherwise unable to be assigned to the variable
resourceUnavailable	The assignment of the value specified by the variable binding's value field to the specified variable requires the allocation of a resource that is presently unavailable
genErr	All other errors

In the second phase, for each variable binding in the request, the named variable is created if necessary, and the specified value is assigned to it. Each of these variable assignments occurs as if simultaneously with respect to all other assignments specified in the same request.

The purpose of the first phase is to ensure the success of the second phase. However, it is still possible that some of these assignments fail in the second phase. If so, all other assignments should be undone, and the Response-PDU is modified to have the value of its error-status field set to commitFailed, and the value of its error-index field set to the index of the failed variable binding.

If and only if it is not possible to undo all the assignments, then the Response-PDU is modified to have the value of its error-status field set to undoFailed, and the value of its error-index field is set to zero.

Finally, the generated Response-PDU is encapsulated into a message and transmitted to the originator of the SetRequest-PDU.

2.4.6   Trap

An SNMPv2-Trap-PDU is generated and transmitted by an SNMP agent when an exceptional situation occurs. See Figure 2.8.

Manager   Network   Agent

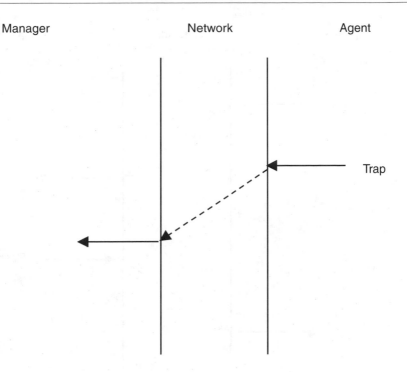

                                                                    Trap

Figure 2.8 Trap Generation.

The destination to which an SNMPv2-Trap-PDU is sent is determined in an implementation-dependent fashion by the SNMPv2 entity. The first two variable bindings in the variable binding list of an SNMPv2-Trap-PDU are sysUpTime.0 and snmpTrapOID.0, respectively. If the OBJECTS clause is present in the invocation of the corresponding NOTIFICATION-TYPE macro, then each corresponding variable, as instantiated by this notification, is copied, in order, to the variable-bindings field. If any additional variables are being included (at the option of the generating SNMPv2 entity), then each is copied to the variable-bindings field.

2.4.7 inform-request

This is also an SNMPv2 protocol operation. An InformRequest-PDU is generated and transmitted at the request of an application in an SNMPv2 entity acting in a manager role that wishes to notify another application (in an SNMPv2 entity also acting in a manager role) of information in an MIB view that is remote to the receiving application. See Figure 2.9.

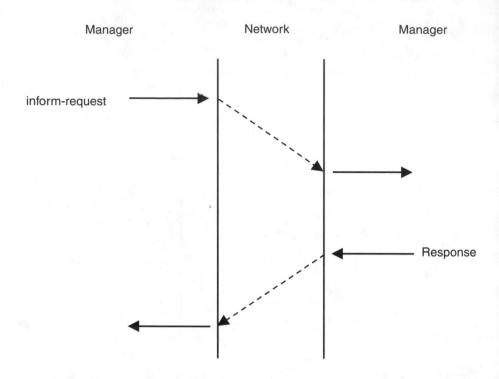

Figure 2.9 Inform-request operation.

The inform-request is sometimes called *confirmed trap* in that the role of this request and its PDU generation are very similar to that of a trap. The destination(s) to which an InformRequest-PDU is sent is specified by the requesting application. The first two variable bindings in the variable-binding list of an InformRequest-PDU are sysUpTime.0 and snmpTrapOID.0, respectively. If the OBJECTS clause is present in the invocation of the corresponding NOTIFICATION-TYPE macro, then each corresponding variable, as instantiated by this notification, is copied, in order, to the variable-bindings field.

Upon receipt of an InformRequest-PDU, the receiving SNMPv2 entity responds to the originator with a Response-PDU. If the generation of the PDU is not successful, an error code may be generated as usual. Otherwise, the receiving SNMPv2 entity performs the following functions.

1. Presents its contents to the appropriate SNMPv2 application;
2. Generates a Response-PDU with the same values in its request-id and variable-bindings fields as the received InformRequest-PDU, with the

value of its error-status field set to noError and the value of its error-index field set to zero;

3. Transmits the generated Response-PDU to the originator of the InformRequest-PDU.

It should be noted that the inform-request could not be used simply as a *confirmed trap*. The inform-request is provided for use in manager-to-manager communication, where one manager has delegated a certain responsibility to another manager. Under these circumstances, manager-A requests manager-B to send InformRequests to notify it of specific conditions that may occur and to retransmit such an inform-request for a defined number of times until the manager-A acknowledges it with a Response-PDU.

2.5 Management Information Base

MIB is a virtual information store of managed objects through which network applications may retrieve information from and exert some form of control over managed devices. MIBs comprise the collected thoughts of experts of the strategy on how to manage a device and on what exactly constitutes the valuable data points needed to monitor and control a device.

This section will introduce a portion of the Internet-standard MIB (MIB-II), specifically system group and interfaces group. SNMP-managed devices typically need to implement at least these groups. Furthermore, understanding of these groups is essential for readers to comprehend the MIB modules that will be introduced in Part Two and Three.

Before getting into details of interested MIB-II groups, let us first examine how to access a conceptual table, a mechanism that is used extensively in defining MIB modules.

2.5.1 Conceptual Table Handling

Conceptual table handling is a mechanism for grouping relevant managed objects. However, this grouping relationship exists only as a characteristic of MIB design, rather than in protocol perspective. The implication is that a special set of procedures have to be followed when accessing such a table.

The definition of the row status textual convention provides a standard way to manipulate conceptual rows. This textual convention defines an object and procedures for dynamic row creation and deletion by management applications using the SNMP protocol. In addition, the row creation mechanism allows applications to learn the agent's notion of appropriate default values for the new row and to use or modify those values. Similarly, the application can discover columns that are not implemented so that it may ignore them and continue to interoperate.

To use this mechanism, a conceptual row should have a status column with a SYNTAX clause value of RowStatus. The status column has six defined values listed in Table 2.13.

2.5.1.1 Conceptual Row Creation

There are four potential interactions when creating a conceptual row: selecting an instance-identifier that is not in use; creating the conceptual row; initializing objects for which the agent does not supply a default; and making the conceptual row available for use by the device.

Selecting an Instance-Identifier

The selection of instance-identifier depends on the design of the conceptual table to be accessed. There are two possible cases as follows:

Table 2.13
Values for RowStatus Object

Value	Explanation	Operations	
		Write	Read
active	The conceptual row is available for use by the managed device	√	√
notInService	The conceptual row exists in the agent but is unavailable for use by the managed device	√	√
notReady	The conceptual row exists in the agent but is missing information necessary to be available for use by the managed device		√
createAndGo	To create a new instance of a conceptual row and to have its status automatically set to active, making it available for use by the managed device	√	
createAndWait	To create a new instance of a conceptual row but not make it available for use by the managed device	√	
destroy	To delete all of the instances associated with an existing conceptual row	√	

- If the instance-identifier is semantically significant (e.g., the destination address of a route) then the instance-identifier should be selected according to the semantics.
- If the instance-identifier is used solely to distinguish conceptual rows, then typically there are two possibilities described as follows.
  - One or more objects provided by the MIB module that defines the conceptual row. These provide assistance in determining an unused instance-identifier, such as the next index object. After each retrieval, the agent modifies the value to the next unassigned index.
  - A pseudo-random number selected by the management station to be used as the index. In the event that this index was already in use and an inconsistentValue was returned in response to the management protocol set operation, the management station should simply select a new pseudo-random number and retry the operation.

Creating the Conceptual Row

Once an unused instance-identifier has been selected, the management station has two choices as follows.

- One-shot creation: To create and activate the conceptual row in one set-request transaction;
- Negotiated creation: To create and activate the conceptual row in a negotiated set of interactions.

Negotiated Conceptual Row Creation   The management station first issues a set-request operation that sets the row status instance to createAndWait. The agent may respond with one of the following:

- wrongValue: Indicating that the agent is unwilling to process the request;
- noError: Indicating that the conceptual row is created. The status column is immediately set to either notInService or notReady, depending on whether there is sufficient information to make the conceptual row available for use by the managed device.

The next step is to initialize nondefaulted objects in the row. The management station issues a get-request to examine all columns in the created

conceptual row. In the response for each column there are three possible out-
comes listed as follows.

- a value: Indicating that some other management station has already
  created this conceptual row. In this case, a new unused instance-
  identifier should be selected;

- noSuchInstance: Indicating that the agent implements the object but
  that the particular instance does not exist;

- noSuchObject: Indicating that the agent does not implement the
  object-type associated with this column.

If the value associated with the status column is notReady, then the management station must first deal with all noSuchInstance columns, if any, until it
becomes notInService.

Subsequently, the management station issues a set-request to set the
status to active. If the agent has sufficient information to make the conceptual
row available for use by the managed device, the management protocol set
operation succeeds with a noError response returned to indicate that the conceptual row has been successfully created. Otherwise, the set operation fails
with an error of inconsistentValue.

One-Shot Conceptual Row Creation   The management station must first
determine the column requirements—i.e., it must determine those columns for
which it must provide values. Depending on the complexity of the table and
the management station's knowledge of the agent's capabilities, this determination can be made locally by the management station. Alternately, the management station issues a management protocol get operation to examine all
columns in the conceptual row that it wishes to create, as discussed in the negotiated creation. Once the column requirements have been determined, a management protocol set operation is issued. This operation also sets the new
instance of the status column to createAndGo.

When the agent processes the set operation, it verifies that it has sufficient
information to make the conceptual row available for use by the managed
device. The information available to the agent is provided by two sources as
follows.

- The variable-binding fields of set-request by management application;

- Implementation-specific defaults supplied by the agent.

If there is sufficient information available, then the conceptual row is created, a noError response is returned, and the status column is set to active.

If there is insufficient information, then the conceptual row is not created, and the set operation fails with an error of inconsistentValue.

2.5.1.2 Conceptual Row Modification

In order for the values of conceptual row objects rather than the status column to be modified, some MIB modules require that the row be taken out of service. This can be done by a management protocol set-request operation that sets the instance of the status column to notInService. If the agent is unwilling to do so, the set operation fails with an error of wrongValue. Otherwise, the conceptual row is taken out of service, and a noError response is returned. It is the responsibility of the DESCRIPTION clause of the status column to indicate under what circumstances the status column should be taken out of service.

Alternatively, a management application may choose to destroy the existing row and create a new one with new values.

2.5.1.3 Conceptual Row Deletion

For deletion of conceptual rows, a management protocol set-request operation is issued that sets the instance of the status column to destroy. This request may be made regardless of the current value of the status column; that is, it is possible to delete conceptual rows that are either notReady, notInService, or active. If the operation succeeds, then all instances associated with the conceptual row are immediately removed.

In summary, conceptual rows can be created either in one-shot or in a step-wise negotiated set of interactions. The latter has the advantage of step-wise error checking; however, it requires more resources.

2.5.2 MIB II

The Internet-standard MIB is the MIB module designed for use with network management protocols in TCP/IP-based internets. The second version of this MIB is defined in RFC 1213 [16], "Management Information Base for Network of TCP/IP-Based Internets: MIB-II." MIB-II is a full Internet standard that conforms to SNMPv1 framework. The interface's group of MIB-II is extended by RFC 1573 [17], "Evolution of the Interfaces Group of MIB-II." The SNMP group is replaced by RFC 1908 [10], "Management Information Base for Version 2 of the Simple Network Management Protocol (SNMPv2)." Both are based on SNMPv2. The structure of MIB-II is depicted in Figure 2.10, which consists of 10 managed object groups that address nearly all aspects of the TCP/IP protocol suite. This book, however, focuses on only two

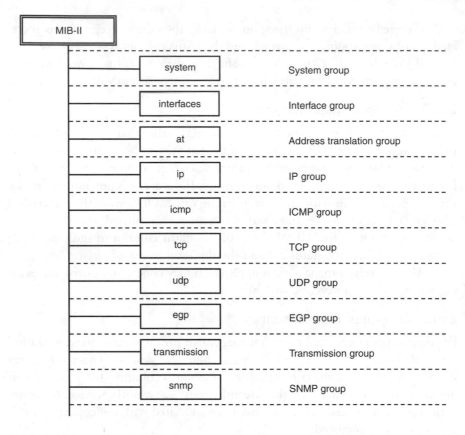

Figure 2.10 Structure of MIB II.

groups, the system group and the interfaces group. These are the groups that usually need to be implemented to work with the ATM-specific MIB modules that will be introduced in Part Two and Part Three.

2.5.2.1 System Group

The system group provides general information for the managed node, which is detailed in Table 2.14. The objects in this group are all scalar ones, reflecting the fact that only one instance of these objects is necessary for a particular managed node. This group should be implemented for every managed node.

Throughout this book, tables are used to describe managed objects. In such tables, the item Prefix represents the prefix for all objects in a group. The actual name of an object in the table is the combination of the Prefix item and the Name item that only shows a short name of an object. For example, the actual object name for "Descr" is "sysDescr," since its prefix is "sys." The OID

Table 2.14
Objects in System Group

Prefix	sys		
OID	{mib-2 1}		
Type of Objects	Name	Definition	Comments
Per System Info	Descr	Description of managed system	
	ObjectID	OID from vendor that identifies the system	
	UpTime	The time (in hundredths of a second) since the system was last reinitialized	
	Contact	Information of contact person for this managed node	
	Name	An administratively assigned textual name for this managed node	
	Location	Physical location of this device	
	Services	A value that indicates the set of services that this entity primarily offers	The value is a sum of individual values, each representing a particular switching function (see Table 2.15 for details)

item shows the OBJECT IDENTIFIER for the root of objects in the group. Note that object names are actually descriptors that are designed to be easily understood by human beings (e.g., the network manger or developers). The guaranteed unique name of an object is the OID.

2.5.2.2 Interfaces Group

The interfaces group of MIB-II is a key component of the core IETF SNMP model of managed systems, which is, as the name suggests, used to manage network interfaces in managed nodes.

In a TCP/IP-based network, an internetwork-layer protocol usually considers any and all protocols it runs over as a single *network interface* layer. This concept is represented in MIB-II by the interfaces group, which defines a generic set of managed objects such that any network interface can be managed in an interface-independent manner through these managed objects. The interfaces group also provides the means for additional managed objects specific to particular types of network interfaces (e.g., a specific medium such as ATM) to be defined as extensions to the interfaces group for media-specific management. Some media-specific extensions that are used to manage ATM networks will be described in Parts Two and Three.

This group contains one scalar object, ifNumber, which represents the number of network interfaces present on this system, and a table, ifTable, in which each interface is represented by a row in the ifTable and identified by a unique value of the ifIndex object. Detailed descriptions of these managed objects are provided by Table 2.16, which can be divided into several categories. Objects in the general info category provide general configuration information for the interface.

The status category lists all the status of the interface. The ifAdminStatus object is a means for conveying an imperative action to the agent. For example, if the value is changed from down to up by a management application, the agent should act accordingly by configuring the interface into the desired state indicated by ifOperStatus. The LastChange object records the time at which the interface entered its current operational state.

Table 2.15
System Services for Each Layer

Layer	Functionality	Value
1	Physical (e.g. repeaters)	1
2	Datalink/subnetwork	2
3	Internet (e.g. IP routers)	4
4	End-to-end (e.g. IP hosts)	8
7	Applications (e.g. mail relays)	64
Note: ATM is a layer 2 protocol in this model.		

Table 2.16
Interface Table

Prefix	if		
OID	{mib-2 2}		
Index	ifIndex		
Type of Objects	Name	Definition	Comments
Index	Index	Value used to identify an interface	
General info	Descr	Description of the interface	
	Type	Type of interface, distinguished according to the physical/link protocol(s) immediately below the network layer in the protocol stack	For media-specific extension
	Mtu	Maximum size of datagram	
	Speed	Estimated current bandwidth of the interface	
	PhysAddress	Media-specific address	
Status	AdminStatus	Desired interface state	
	OperStatus	Current operational state of the interface	
	LastChange	Time when the interface entered its current operational state	
Counters for incoming packets	InOctets	Total number of octets received on the interface	
	InUcastPkts	Subnetwork-unicast packets delivered to a higher-layer protocol	
	InNUcastPkts	Broadcast/multicast packets delivered to a higher-layer protocol	Deprecated*
	InDiscards	Discarded incoming packet due to resource limitations	
	InErrors	Incoming packets discarded due to errors	

Table 2.16 (continued)

Prefix	if		
OID	{mib-2 2}		
Index	ifIndex		
Type of Objects	Name	Definition	Comments
	InUnknownProtos	Incoming packets discarded due to unknown or unsupported protocol	
Counters for outgoing Packets	OutOctets	Total number of octets transmitted out of the interface	
	OutUcastPkts	Total number of unicast packets that higher-level protocols requested be transmitted	
	OutNUcastPkts	Total number of non-unicast packets that higher-level protocols requested be transmitted	Deprecated*
	OutDiscards	Number of output packets discarded due to resource limitations	
	OutErrors	Number of outbound packets that could not be transmitted because of errors	
Misc	OutQLen	Length of the output packet queue	Deprecated*
	Specific	Media-specific MIB pointer	Deprecated*

* Deprecated in RFC 1573.

The counters categories provide performance statistics for this interface in both incoming and outgoing directions.

With the objects in this group, a network management application can monitor the operation and easily find most of the problems associated with network interfaces.

2.5.2.3 Interfaces Group Evolution

Experience in defining media-specific MIB modules has shown that the model defined by MIB-II is too simplistic and/or static for some types of media-specific management. To address this problem and others, RFC 1573 was proposed to evolve the interfaces group. As of this writing, a new interface MIB, RFC 2233, has just been published. Nevertheless, we will focus on RFC 1573, as many MIBs introduced in later chapters refer to this MIB.

One of the major contributions of RFC 1573 is that it allows a complicated interface relationship to be represented. The original model of the interfaces group was that each network interface represented a complete interface stack that spanned from immediately underneath the internet layer down to the physical layer. It was unable to represent upward and downward multiplexing of the sublayers. The evolution permits each sublayer to be defined as a network interface and defines a new table, the ifStackTable, to represent how such interfaces are layered.

RFC 1573 also makes possible the dynamic addition/removal of interfaces. The original MIB-II description of ifIndex not only constrains its value to start at one and be less or equal to the number of interfaces but also requires that the value for a particular interface must not change except at a restart of the managed agent. This represents a problem for agents that can have a network interface dynamically removed—e.g., when an interface other than the one with the highest ifIndex value is unconfigured. The evolution solves this by allowing the value of ifIndex to be greater than the current number of interfaces, although it still recommends that ifIndex values be assigned contiguously starting at one.

Structure of the New Interfaces Group

The new MIB consists of five tables, as well as definitions for link traps:

- ifTable: This table is the ifTable from MIB-II.

- ifXTable: This table contains objects that have been added to interface MIB as a result of the interface evolution effort, or replacements for objects of the original, MIB-II, ifTable that were deprecated because the semantics of said objects have significantly changed.

- ifStackTable: This table contains objects that define the relationships among the sub-layers of an interface.

- ifTestTable: This table contains objects that are used to perform tests on interfaces. This table is a generic table. Accordingly, the designers of media-specific MIBs must define exactly how this table applies to their specific MIBs.

- ifRcvAddressTable: This table contains objects that are used to define the media-level addresses that this interface will receive. Again, this table is a generic table that requires the designers of media-specific MIBs to define exactly how the table applies to their specific MIBs.

In addition, a new textual convention, IANAifType, has been defined for the enumerated values of ifType. This allows new ifType values to be assigned by the *Internet Assigned Number Authority* (IANA). The new MIB includes an initial version of IANAifType that will be updated and periodically reissued by the IANA.

ifTable.   This table is generally the same as the original MIB-II ifTable except for refinements of some objects. By making use of SNMPv2 capability to define overlapping MIB groups for conformance purposes, objects in this table are further divided into three groups to support bit- and character-oriented interfaces. An ifGeneralGroup applies to all network interfaces; an ifPacket-Group group contains those objects applicable to all packet-based interfaces; and an ifFixedLengthGroup contains objects applicable to interfaces that transmit data in fixed-length transmission units, including character-oriented interfaces.

ifXTable.   This table contains additional objects for the interfaces table as listed in Table 2.17. It is an extension of the MIB-II ifTable. As such, it augments the ifEntry. The most notable change is the definition of a set of objects for high-speed interface, such as 64-bit counters. In addition, the table contains definitions of broadcast and multicast objects that replace the InNU-castPkts/OutNUcastPkts in the original MIB-II interfaces group.

ifStackTable.   The table provides information on the relationships between the multiple sublayers of network interfaces, which is shown in Table 2.18. In particular, it contains information on which sublayers run "on top of" other sublayers. Each sublayer corresponds to a conceptual row in the ifTable. With this table, a complicated relationship between sublayers of an interface can be represented.

An ATM switch that supports virtual path tunneling service may maintain several logical port interfaces over a physical interface. Please refer to Chapter 11 for more details on this service. Figure 2.11 shows an example of an ATM interface that supports three logical ports. The ifStackTable can be used to describe the relationship between the interfaces involved. Table 2.19 lists relevant entries that are used.

Table 2.17
Interface Extension Table

Prefix	if		
OID	{ ifMIBObjects 1 }		
Index	AUGMENTS ifEntry, i.e., indexed by ifIndex		

Type of objects	Name	Definition	Comments
Multicast and broadcast counters	InMulticastPkts	Number of multicast packets delivered to a higher-layer protocol	
	InBroadcastPkts	Number of packets delivered to a higher layer	
	OutMulticastPkts	Number of broadcast packets that higher-level protocols requested be transmitted	
	OutBroadcastPkts	Number of broadcast packets that higher-level protocols requested be transmitted	
High Capacity counters	HCInOctets	Number of octets received on the interface	64-bit version of corresponding objects in ifTable
	HCInUcastPkts	Number of received Unicast packets, delivered by this sublayer to a higher (sub)layer,	
	HCInMulticastPkts	Number of multicast packets delivered to a higher-layer protocol	
	HCInBroadcastPkts	Number of broadcast packets delivered to a higher-layer protocol	
	HCOutOctets	Number of octets transmitted out of the interface	

Table 2.17 (continued)

Prefix	if		
OID	{ ifMIBObjects 1 }		
Index	AUGMENTS ifEntry, i.e., indexed by ifIndex		
	HCOutUcastPkts	Number of unicast packets that higher-level protocols requested be transmitted out of the interface	
	HCOutMulticastPkts	Number of multicast packets that higher-level protocols requested be transmitted out of the interface	
	HCOutBroadcastPkts	Number of broadcast packets that higher-level protocols requested be transmitted out of the interface	
	HighSpeed	An estimate of the interface's current bandwidth in units of 1,000,000 bits per second	Also for high-speed interface
Misc	Name	Textual name of the interface	
	LinkUpDownTrap Enable	Indicating whether linkUp/linkDown traps should be generated for this interface	See link traps for details
	PromiscuousMode	Indicating whether the station accepts all packets/frames transmitted on the media	
	ConnectorPresent	Indicating whether the interface sublayer has a physical connector	

Table 2.18
Interface Stack Table

Prefix	ifStack		
OID	{ ifMIBObjects 2 }		
Index	ifStackHigherLayer, ifStackLowerLayer		
Type of objects	Name	Definition	Comments
	HigherLayer	ifIndex corresponding to the higher sublayer	
	LowerLayer	ifIndex corresponding to the lower sublayer	
	Status	RowStatus for this row	

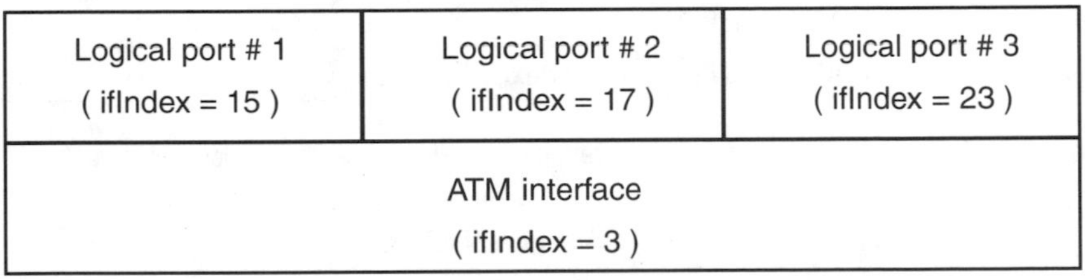

Logical port # 1 (ifIndex = 15)	Logical port # 2 (ifIndex = 17)	Logical port # 3 (ifIndex = 23)
ATM interface (ifIndex = 3)		

Figure 2.11 ATM logical interface example.

ifTestTable. This table contains objects that allow a network manager to instruct an agent to test an interface for various faults. Tests for an interface are defined in the media-specific MIB for that interface. After invoking a test, the object ifTestResult can be read to determine the outcome. The object ifTest-Code can be used to provide further test-specific or interface-specific (or enterprise-specific) information concerning the outcome of the test. Only one test can be in progress on each interface at any one time. To ensure this, the ifTestId object with syntax TestAndIncr is used. This object provides a semaphore-like mechanism required for multiple management applications to access the same row in the table concurrently. Table 2.20 lists all the objects defined in the ifTestTable.

Table 2.19
ifStackTable Entries for the ATM Logical Interface Example

HigherLayer	LowerLayer	RowStatus
15	3	active(1)
17	3	active(1)
23	3	active(1)

Table 2.20
Interface Test Table

Prefix	ifTest		
OID	{ ifMIBObjects 3 }		
Index	AUGMENTS ifEntry, i.e., indexed by ifIndex		
Type of objects	Name	Definition	Comments
	Id	Identifies the current invocation of the interface's test	
	Status	Indicates whether the manager currently has the necessary "ownership" required to invoke a test on this interface	
	Type	Type of the test (e.g., loopback test)	
	Result	Result of the most recently requested test	
	Code	Specific information on the test result (e.g., error code)	
	Owner	Current owner of this test	

ifRcvAddressTable. This table contains information of addresses for which the system will accept packets/frames on the particular interface identified by the index value ifIndex. The objects in this table are listed in Table 2.21.

Table 2.21
Receive Address Table

Prefix	ifRcvAddress		
OID	{ ifMIBObjects 4 }		
Index	ifIndex, ifRcvAddressAddress		
Type of objects	Name	Definition	Comments
	Address	An address for which the system will accept packets/frames on this entry's interface	
	Type	Memory type to store the address (e.g., volatile, nonVolatile)	
	Status	Row status	

Link Traps. RFC 1573 also defines two traps for interfaces—link-up trap and link down trap—to notify management applications of the changes of status of an interface. Three objects, ifIndex, ifAdminStatus, and ifOperStatus are used to identify the interface that generates the trap and to report its status.

The ifLinkUpDownTrapEnable object in the ifXTable provides a mechanism for management stations to control the generation of link traps. This object allows managers to limit generation of traps to just the sublayers of interest. This mechanism is especially useful in a multiple-layer interface in which link state changes may propagate from bottom to top, or vice versa. Among conditions that cause these state changes the one that is of most interest to network managers usually occurs at the lowest level of an interface stack. The generation of traps from other sublayers should be disabled by setting the ifLinkUpDownTrapEnable object to disabled(2).

```
linkDown NOTIFICATION-TYPE
    OBJECTS { ifIndex, ifAdminStatus, ifOperStatus }
    STATUS  current
    DESCRIPTION
        A linkDown trap signifies that the SNMPv2
        entity, acting in an agent role, has
        detected that the ifOperStatus object for
```

```
            one of its communication links
            is about to transition into the down state.
        ::= { snmpTraps 3 }


    LinkUp NOTIFICATION-TYPE
        OBJECTS { ifIndex, ifAdminStatus, ifOperStatus }
        STATUS   current
        DESCRIPTION
            A LinkUp trap signifies that the SNMPv2
            entity, acting in an agent role, has
            detected that the ifOperStatus object for
            one of its communication links has
            transitioned out of the down state.
        ::= { snmpTraps 4 }
```

2.6   Coexistence of SNMPv1 and SNMPv2

Approaches for the coexistence of SNMPv1 and SNMPv2 have been described in RFC 1908, "Coexistence between Version 1 and Version 2 of Internet-standard Network Management Framework." It consists of two parts, management information and protocol operations.

2.6.1   MIB Modules

2.6.1.1   V1-to-V2

SNMPv2 was carefully designed to minimize interworking problems with SNMPv1. The SNMPv2 approach toward describing collections of managed objects is nearly a proper superset of the approach defined in SNMPv1. MIB modules defined using the SNMPv1 current framework may continue to be used with SNMPv2. For example, an implementor can take an SNMPv1 product, add support for SNMPv2, and not have to change any of the code that implements MIB support in the product. This means that any MIB module compliant to SNMPv1's SMI will work with an SNMPv2 product. However, for the MIB modules to conform to the SNMPv2 framework, a few changes are required; these changes are detailed in RFC 1908 [10].

2.6.1.2   V2-to-V1

SNMPv2 was proposed to replace SNMPv1. Unfortunately, this goal has not yet been achieved. Most network management systems are still SNMPv1 because it is too good to abandon. Furthermore, many MIB modules that are compliant to SNMPv2 have been used in SNMPv1 products. Thus, it is sometimes necessary to use MIB modules defined by SNMPv2 SMI in SNMPv1 products. Fortunately, there are very few features in SNMPv2's SMI that are incompatible with SNMPv1. As such, SNMPv2 MIB modules can be specified, with some care, in a manner that is both compliant to SNMPv2 SMI and semantically identical to the peer SNMPv1 definitions. The conversion of an SNMPv2 MIB module to its SNMPv1 counterpart can be done automatically. In fact, the *Simple Times* [18], a publication devoted to the promotion of the SNMP, has introduced an e-mail address for that purpose. Any SNMPv2 MIB module sent to the following address will be converted to SNMPv1.

mib-v2tov1@simple-times.org

2.6.2 Protocol Operations

This section examines mechanisms for the co-existence of SNMPv1 and SNMPv2 products. Two transition mechanisms are recommended in the SNMPv2 coexistence document: proxy agents and bilingual managers; also recommended is the bilingual agent, which is an industrial solution.

2.6.2.1 Proxy Agent

The proxy agent translates SNMPv2 packets from management stations into SNMPv1 packets to be sent to agents, and SNMPv1 replies from agents into SNMPv2 packets to be sent to management stations. The rules for translation in the proxy agent are described in the following sections.

V2 to V1

- If a GetRequest-PDU, GetNextRequest-PDU, or SetRequest-PDU is received, then it is passed unaltered by the proxy agent.
- If a GetBulkRequest-PDU is received, the proxy agent sets the nonrepeaters and max-repetitions fields to zero and sets the tag of the PDU to GetNextRequest-PDU.

V1 to V2

- If a GetResponse-PDU is received, then it is passed unaltered by the proxy agent, even though an SNMPv2 entity will never generate a Response-PDU with a error-status field having a value of noSuchName, badValue, or readOnly. This allows the SNMPv2 manager to interpret the response correctly. However, if the error-status field has a value of tooBig, the proxy agent will remove the contents of the variable-bindings field before propagating the response.

- If a Trap-PDU is received, then it is mapped into an SNMPv2-Trap-PDU. This is done by prepending onto the variable-bindings field two new bindings: sysUpTime.0 and snmpTrapOID.0. Then, snmpTrap-Enterprise.0, a new binding that takes its value from the enterprise field of the Trap-PDU is appended onto the variable-bindings field. The destinations for the SNMPv2-Trap-PDU are determined in an implementation-dependent fashion by the proxy agent.

Proxy agents add complexity that is visible to a network manager. When faced with lack of response from an agent, a network manager using a proxy agent will need to determine if the proxy agent is running, rather than being able to simply trust the response from the management tool. The proxy agent also adds a level of hierarchy to the configuration of the network management system that adds complexity. If product developers need to use a proxy agent solution, they should work hard to shield this complexity from the user so that their tool is easily trusted.

2.6.2.2 Bilingual Manager

The bilingual manager solution is implemented by providing a management station with the capability to send either SNMPv1 or SNMPv2 packets. The choice of protocol is a local configuration matter on the management station and is typically made on a per-device basis, although automatic configuration is possible for a small system.

This approach is nearly seamless to the network management user. Most operations are performed in exactly the same way regardless of which protocol is being used. Those hosts that are configured to use SNMPv2 will perform management operations faster and have some additional features available to them.

To provide transparency to management applications, the entity acting in a manager role must map operations as if it were acting as a proxy agent. When designing an SNMPv2-enabled application, it would be helpful to consult the rules for proxy agent behavior for ideas on how to retain coexistence with SNMPv1. For example, these rules describe how an application can translate a get-bulk PDU into a get-next PDU for transmission to an SNMPv1 agent.

2.6.2.3 Bilingual Agent

The third mechanism that may be used is the bilingual agent [19–20] . This is implemented by allowing an agent to respond to either SNMPv1 or SNMPv2 requests. It is fairly simple to implement and does not add a lot of cost to the agent. However, this approach is not explicitly part of the transition plan, as it places an additional burden on the agent.

Many managed node vendors choose to use this approach due to ease of implementation, simplicity of installation and configuration, and low additional cost. The additional cost of implementing both protocols is worth the advantage of retaining backward compatibility with older management stations.

In summary, this chapter introduces the fundamentals of SNMP. For a detailed description of this framework, please refer to [21–22].

References

[1] Rose, M., and K. McCloghrie, "Structure and Identification of Management Information for TCP/IP-based Internets," RFC 1155, May 1990.

[2] Rose, M., and K. McCloghrie, editors, "Concise MIB Definitions," RFC 1212, March 1991.

[3] Case, J., M. Fedor, M. Schoffstall, and J. Davin, "Simple Network Management Protocol," RFC 1157, May 1990.

[4] SNMPv2 Working Group, J. Case, K. McCloghrie, M. Rose, and S. Waldbusser, "Structure of Management Information for Version 2 of the Simple Network Management Protocol (SNMPv2)," RFC 1902, January 1996.

[5] SNMPv2 Working Group, J. Case, K. McCloghrie, M. Rose, and S. Waldbusser, "Textual Conventions for Version 2 of the Simple Network Management Protocol (SNMPv2)," RFC 1903, January 1996.

[6] SNMPv2 Working Group, J. Case, K. McCloghrie, M. Rose, and S. Waldbusser, "Conformance Statements for Version 2 of the Simple Network Management Protocol (SNMPv2)," RFC 1904, January 1996.

[7] SNMPv2 Working Group, J. Case, K. McCloghrie, M. Rose, and S. Waldbusser, "Protocol Operations for Version 2 of the Simple Network Management Protocol (SNMPv2)," RFC 1905, January 1996.

[8] SNMPv2 Working Group, J. Case, K. McCloghrie, M. Rose, and S. Waldbusser, "Transport Mappings for Version 2 of the Simple Network Management Protocol (SNMPv2)," RFC 1906, January 1996.

[9] SNMPv2 Working Group, J. Case, K. McCloghrie, M. Rose, and S. Waldbusser, "Management Information Base for Version 2 of the Simple Network Management Protocol (SNMPv2)," RFC 1907, January 1996.

[10] SNMPv2 Working Group, J. Case, K. McCloghrie, M. Rose, and S. Waldbusser, "Coexistence Between Version 1 and Version 2 of the Internet-Standard Network Management Framework," RFC 1908, January 1996.

[11] Information processing systems—Open Systems Interconnection—Specification of Abstract Syntax Notation One (ASN.1), International Organization for Standardization. International Standard 8824, December, 1987.

[12] Information processing systems—Open Systems Interconnection—Specification of Basic Encoding Rules for Abstract Syntax Notation One (ASN.1), International Organization for Standardization. International Standard 8825, December, 1987.

[13] Network Working Group, M. Rose, "A Convention for Defining Traps for Use With the SNMP," RFC 1215, March 1991.

[14] McCloghrie, K., "Security and Protocols," *The Simple Times,* Volume 2, Number 5, September/October, 1993.

[15] Waldbusser, S. L., "Applications Stand To Benefit From SMP," *The Simple Times,* Volume 1, Number 4, September/October, 1992.

[16] McCloghrie, K., and M. Rose, editors, "Management Information Base for Network Management of TCP/IP-Based Internets: MIB-II," STD 17, RFC 1213, March 1991.

[17] Network Working Group, K. McCloghrie, and F. Kastenholz "Evolution of the Interfaces Group of MIB-II," RFC 1573, January 1994.

[18] *The Simple Times,* http://www.simple-times.org/.

[19] Waldbusser, S. L., " Hints on Coexistence and Transition From SNMP to SNMPv2," *The Simple Times,* Volume 2, Number 2, March/April, 1993.

[20] Network Working Group, B. Wijnen, and D. Levi, "V2ToV1 Mapping SNMPv2 onto SNMPv1 Within a Bi-lingual SNMP Agent," RFC 2089, January 1997.

[21] Rose, M. T., *The Simple Book: An Introduction to Internet Management,* Second Edition, Englewood Cliffs, New Jersey: Prentice-Hall, 1994.

[22] Stallings, W., *SNMP, SNMPv2 and CMIP: The Practical Guide to Network Management Standards,* Reading, MA: Addison-Wesley Publishing Co, Inc., 1993.

3

Asynchronous Transfer Mode

3.1 Introduction

ATM, also known as cell relay, is the technology that has been chosen for *broadband integrated services digital network* (B-ISDN). B-ISDN standards are defined by the Telecommunication Standardization Sector of the *International Telecommunication Union* (ITU-T). The ATM Forum, an international non-profit organization, is also very active in defining industrial standards to accelerate the use of ATM products and services. For a list of ATM standards from ITU-T and the ATM Forum, please refer to Appendix B.

Figure 3.1 depicts the structure and interfaces of ATM networks defined by the ATM Forum. ATM can be used for both public carrier networks and private networks. In recognition of this fact, two distinct forms of ATM *user network interface* (UNI) are defined. Public UNI is typically used to interconnect ATM user equipment such as *terminal equipment* (TE) or a private network with an ATM switch deployed in a public service provider's network. Note that the TE may be a host with an ATM card that runs applications over ATM or a router that forwards IP packets to a LAN through the ATM network. Private UNI is used to interconnect an ATM user with an ATM switch that is managed as part of the same corporate network. Similarly, there are two forms of *network node interface* (NNI), public NNI for interconnecting public networks and PNNI for private networks. The *broadband-intercarrier interface* (B-ICI) is defined for interworking between carriers.

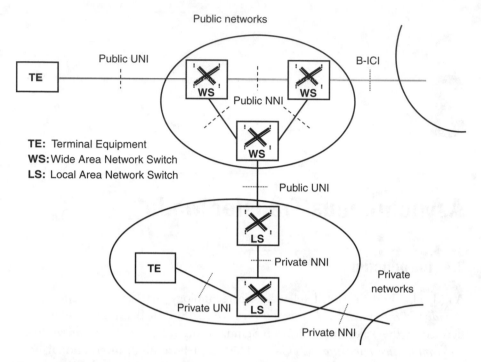

Figure 3.1 ATM network structure.

3.2 Protocol Reference Model

Figure 3.2 depicts the B-ISDN protocol reference model [1], which contains three separate planes. The user plane is for the transfer of end-user information through the network. This plane is concerned primarily with the ATM layer and the physical layer—i.e., the layers most relevant to accomplishing cell relay services in an ATM network. User plane functions will be detailed in Sections 3.3 and 3.4.

The control plane deals with call control and connection control functions (e.g., signaling). Signaling is the automated message exchange mechanism by which switched network connections can be established or released on demand. The control plane shares the ATM layer and physical layer facilities with the user plane. ATM is a connection-oriented transfer mechanism; this means that every switched connection within the ATM layer must first be allocated an identifier via signaling procedures and that resources must be reserved in the network. Typically, the control plane is responsible for the set-up and tear-down of *switched virtual connections* (SVCs) within an ATM network. A

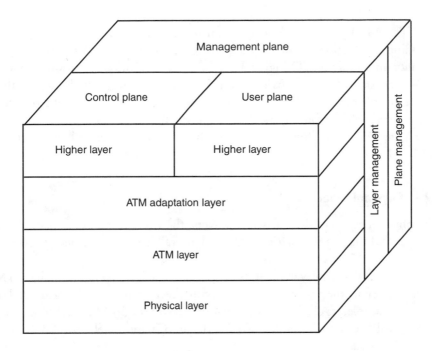

Figure 3.2 B-ISDN protocol reference model. (*Source: ITU-T Recommendations, I.321, Figure 1.*)

detailed description of control plane functions is outside the scope of this book. However, PNNI signaling will be covered briefly in Chapter 8 to introduce PNNI management.

The management plane provides functions for operations and management, as well as the capability to exchange information between the user plane and the control plane. The management plane performs two primary functions, layer management for layer-specific functions, such as detecting failures and protocol abnormalities, and plane management for managing and coordinating functions related to the overall ATM architecture. In particular, layer management provides management functions relating to resources and parameters residing in its protocol entities, such as meta-signaling and the OAM information flows. Meta-signaling is a separate information channel used to set up signaling channels. OAM flows are used to monitor network performance and to detect errors for fault management, which will be examined in Chapter 4. The architecture of the management plane functions will also be introduced in Chapter 4. The management plane is responsible for the set-up and release of permanent virtual connections, which will be explained in Chapter 6.

The ATM layer offers a flexible transfer capability common to all services, including connection-oriented and connectionless services. Additional functionality on top of the ATM layer is provided to accommodate various services. ATM networks do not provide extensive error control, which means that higher layers are responsible for all retransmission. The ATM protocol reference model includes four layers. The detailed structure of this model is shown in Figure 3.3. The functions for each layer are described in Sections 3.2.1–3.2.4.

3.2.1 Physical Layer

The physical layer consists of two sublayers: the *transmission convergence* (TC) sublayer and the *physical medium* (PM) sublayer. They are described as follows.

- TC: This sublayer is responsible for embedding cells from the ATM layer in the transmission frames of the transport medium used. For example, if ATM cells are transmitted over the 34 Mbps E3 links, they will need to be embedded in the information fields of E3 frames.

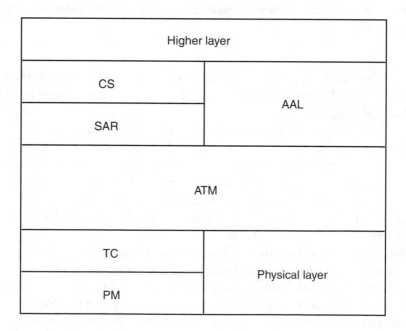

Figure 3.3 Structure of layer functions.

Where cells are carried directly without a transmission frame, this is not necessary. This sublayer consists of the following functions:

- Transmission frame generation/recovery: Generating and maintaining the required physical layer frame structure appropriate for a given data rate;

- Transmission frame adaptation: Packaging incoming ATM layer cells into frames;

- Cell delineation: Maintaining the cell boundaries so that cells may be recovered after descrambling at their destination;

- HEC sequence generation/cell header verification: Generating and checking the *header error-control* (HEC) code used to protect cell headers;

- Cell rate decoupling: Insertion and suppression of idle cells in order to adapt the rate of valid ATM cells to the payload capacity of the transmission system in use.

- PM: This sublayer includes only PM-dependent functions. Its specification will therefore depend on the medium used. Possibilities include fiber optics, *unshielded twisted pair* (UTP), and radio, to name a few. One function that is common to all media is bit timing. This sublayer is responsible for transmitting/receiving a continuous flow of bits with associated timing information to synchronize transmission and reception.

3.2.2 ATM Layer

The ATM layer functions, which are independent of the PM in use, include the following:

- Cell multiplexing and demultiplexing: Multiplexing and demultiplexing so that all ATM layer cells can be carried over a physical layer medium;

- Cell *virtual path identifier* (VPI)/*virtual channel identifier* (VCI) translation: Converting the VPI and VCI of incoming cells into the ones that are required for the next hop at a switch or a cross-connect during switching, as VCI/VPI only has local significance;

- Cell header generation/extraction: Appending cell headers to user data from the AAL layer for physical layer transmission and extracting user data from cells received. This function may also include the translation of an address to VPI/VCI;

- *Generic flow control* (GFC): Used to control cell transfer in a shared PM by several subscribers; one possible use is to resolve multiple access contention among multiple users at the UNI.

3.2.3 ATM Adaptation Layer

The *ATM adaption layer* (AAL) adapts the services provided by the ATM layer to support the functions required by the next higher layer, for example, the *Internet protocol* (IP). The functions performed in the AAL layer depend on the higher-layer requirements. Therefore, the AAL itself is service-dependent. It consists of two sublayers: the *segmentation and reassembly* (SAR) sublayer and the *convergence sublayer* (CS). They are described as follows.

- SAR: This sublayer is responsible for the segmentation of higher-layer data into a size suitable for an ATM cell information field and the reassembly of the contents of ATM cell information fields into higher-layer information.
- CS: This sublayer provides the *ATM adaptation layer* (AAL) service at the AAL-*service access point* (SAP).

3.2.4 Higher Layer

Higher layers make use of ATM services. Typical higher layer services are frame relay, B-ISDN Signaling, LANE, and TCP/IP.

3.3 ATM Layer

ATM is a connection-oriented packet switching technology. Like X.25, ATM utilizes virtual connections to carry user data over a network. The *virtual channel* (VC) in ATM is similar to the virtual circuit in X.25. However, ATM differs from X.25 in that it uses a two-level virtual connection technology by introducing another virtual connection level termed *virtual path* (VP). Another key difference is that the X.25 packet size is variable, whereas the ATM cell size is fixed, allowing for efficient (hardware) switching.

3.3.1 ATM Transport Network

The ATM transport network [2] is structured as two layers, namely the ATM layer and the physical layer, as shown in Figure 3.4.

The transport functions of the physical layer are subdivided into three levels: the transmission path level, the digital section level, and the regenerator section level. The transmission path extends between network elements that assemble and disassemble the payload of a transmission system. Cell delineation and HEC functions are required at the end point of each transmission path. The digital section extends between network elements that assemble and disassemble a continuous bit or byte stream. The lowest level, the regenerator section level, is a portion of digital section. Note that the above model is only of logical importance. Hence it does not necessarily mean that there is physical equipment corresponding to each level. For example, an ATM card contains all three levels, while SDH line termination equipment only supports two lowest level functions.

The transport functions of the ATM layer are subdivided into two levels, the VC level and the VP level. The hierarchical layer-to-layer relationship of the ATM transport network is shown in Figure 3.5.

		Higher layer	
ATM transport network	ATM layer	VC level
		VP level
	Physical layer	Transmission path level
		Digital section level
		Regenerator section level

Figure 3.4 Hierarchy of the ATM transport network. (*Source: ITU-T Recommendations, I.311, Figure 1.*)

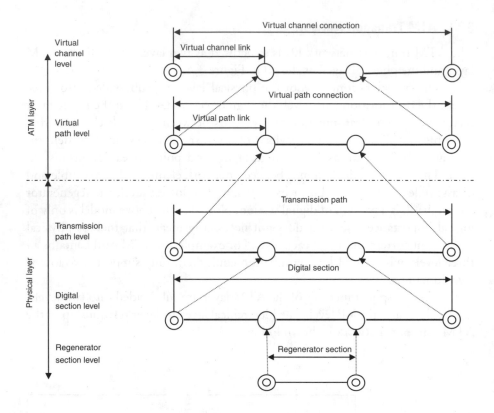

Figure 3.5 Hierarchical layer-to-layer relationship. (*Source: ITU-T Recommendations, I.311, Figure 1.*).

The VP concept was developed in response to a trend in high-speed networking in which the control cost of the network is becoming an increasingly higher portion of the overall network cost. By grouping connections that share common paths through the network into a single unit, network management applications can deal with a small number of groups of connections instead of a large number of individual connections, resulting in increased network performance and reliability. In certain cases, the processing and connection time can also be reduced. For example, by reserving capacity on a *VP connection* (VPC), new *VC connections* (VCCs) can be established by executing simple control functions at the end points of the VPC; no processing is required at transit nodes. Furthermore, the VP technique is very useful in defining closed user groups or virtual private networks, with which network resources are provided to users on demand by public carriers in such a manner that the users view this partition of the network as a private network. Within such a network,

ATM VCCs are tunneled transparently over public networks using VPs. Please refer to Chapter 11 for details.

3.3.2 Virtual Connections

Before getting into details of VP/VC operations, let us examine related terminologies [2] that are used in ITU-T standards for VC/VP levels:

- *VC:* A VC is a generic term used to describe a communications capability for the transport of ATM cells.
- *VCI:* A VCI identifies a *VC link* (VCL) for a given *VP link* (VPL);
- *VCL:* A VCL is a capability for the transport of ATM cells between two adjacent ATM peer entities where the VCI value is translated. A VCL for a given VPC is identified by a VCI.
- *VCC endpoint:* A VCC endpoint is the point where the cell information field is exchanged between the ATM layer and the user of the ATM layer service (e.g, the ATM adaptation layer).
- *VCC:* A VCC is a concatenation of VCLs that extends between two VCC endpoints or, in the case of point-to-multipoint arrangements, more than two VCC endpoints.
- *VP:* A VP is a generic term for a bundle of VCLs; all the VCLs in a bundle have the same endpoints.
- *VPI:* A VPI identifies a group of VCLs, at a given reference point, that share the same VPL.
- *VPL:* A VPL is a capability for the transport of ATM cells between two adjacent ATM peer entities where the VPI value is translated. A VPL is originated or terminated by the assignment or removal of the VPI value.
- *VPC endpoint:* A VPC endpoint is the point at which the VCIs are originated, translated, or terminated.
- *VPC :* A VPC is a concatenation of VPLs that extends between two VPC endpoints or, in the case of point-to-multipoint arrangements, more than two VPC endpoints.

Note that VPI/VCIs are only of local significance and therefore can only be used to identify virtual links. When a VPI/VCI is assigned, the same value is assigned for both directions of transmission. However, the characteristics of the

VPC/VCC in both directions may be the same or different. In some applications, the bandwidth for one direction may be equal to zero.

In contrast to traditional packet-switched networks, ATM networks employ a mechanism to provide guaranteed QoS by reserving network resources. Although this technique existed already in X.25, it was never widely implemented. Each VPC/VCC is associated with a set of QoS parameters such as cell loss ratio, cell delay, and cell delay variation. ATM networks are committed to provide the negotiated QoS parameters as long as the user of the VPC/VCC complies with the negotiated traffic contract. Both the QoS and traffic contract are negotiated between a user and the network for each VPC/VCC when the connection is set up. Input cells from the user to the network are monitored to ensure that the negotiated traffic parameters are not violated. Please refer to Chapter 12 for more details on traffic management.

3.3.2.1 VC / VP Switching

ITU-T has defined six types of switching fabrics that are able to switch virtual connections within an ATM network:

- *VP cross-connect:* A VP cross-connect is a network element that connects VPLs; it translates only VPI values and is directed by management plane functions and not by control plane functions.

- *VC cross-connect:* A VC cross-connect is a network element that connects VCLs; it terminates VPCs and translates VCI values and is directed by management plane functions and not by control plane functions.

- *VP-VC cross-connect:* A VP-VC cross-connect is a network element that acts both as a VP cross-connect and as a VC cross-connect. It is directed by management plane functions and not by control plane functions.

- *VP switch:* A VP switch is a network element that connects VPLs; it translates only VPI values and is directed by control plane functions.

- *VC switch:* A VC switch is a network element that connects VCLs; it terminates VPCs and translates VCI values and is directed by control plane functions.

- *VP-VC switch:* A VP-VC switch is a network element that acts both as a VP switch and as a VC switch. It is directed by control plane functions.

In terms of switching functionality, a cross-connect and a switch are identical. The only difference is that a cross-connect is directed by management plane functions, while a switch is directed by control plane functions. Furthermore, the VP/VC cross-connect and the VP/VC switch in an ATM switch may share the same physical switching fabric. Typically, the virtual connections that are set up by management plane functions are termed semipermanent/ permanent virtual connections; while the connections by control plane functions (e.g., signaling) are referred to as SVCs, reflecting the dynamic nature of the connections.

Figures 3.6 and 3.7 depict the operation of VP switching and VC switching. The VPI/VCI in a switch is translated from input cell stream to output cell stream by a look-up table. The input parameters include the input port number and the VPI/VCI value from an incoming cell. The output parameters are the outgoing VCI/VPI and output port number for the next hop. Table 3.1 is an

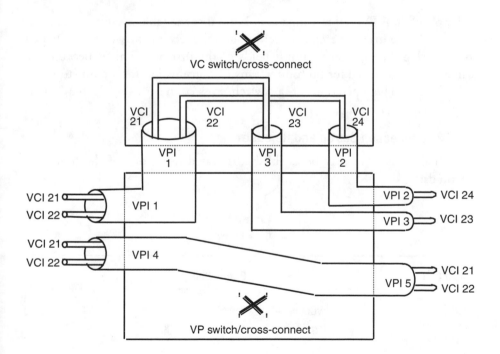

Figure 3.6 VC and VP switching.

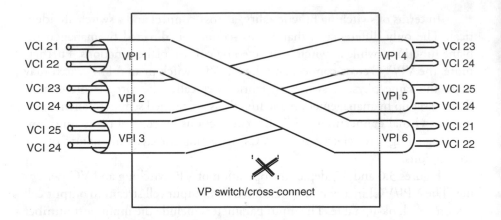

Figure 3.7 VP switching. (*Source: ITU-T Recommendations, I.311, Figure 4.*)

example of an ATM cell routing table. Note that the table is symmetrical, i.e., there are two entries for each connection, and the second entry can be obtained by reversing the input and output fields of the first one. This is because the same value is assigned for both directions of transmission for a point-to-point VPC/VCC. The operation of a switch/cross-connect using Table 3.1 is depicted in Figure 3.8.

3.3.2.2 Connection Set-up and Release

A virtual connection may be established/released using one of the following methods:

Table 3.1
Cell Routing Table

Input Cell			Output Cell		
Port	VPI	VCI	Port	VPI	VCI
1	20	29	2	30	45
2	30	45	1	20	29
1	40	64	3	50	10
3	50	10	1	40	64

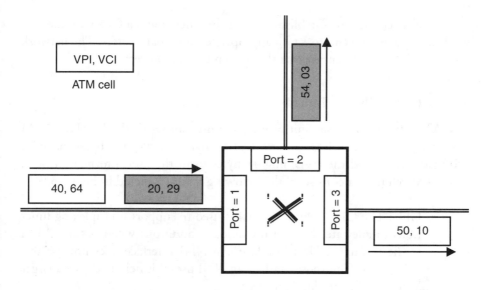

VP-VC switch / Cross-connect

Figure 3.8 Switching fabric operation.

- Management plane functions: A VPC/VCC may be prereserved with the network as in the case of a permanent virtual connection or semi-permanent connection.

- Meta-signalling: A VCC may be set-up via a meta-signalling procedure across a meta-signalling VC that is used to establish, check, and release point-to-point and selective broadcast signaling VCs. It is bi-directional and permanent. B-ISDN meta-signalling protocol is defined in ITU-T Q.2130.

- User-to-network signaling procedures: A VPC/VCC connection maybe set up as a result of a user-to-network signaling procedure that is performed between terminal and network, across a UNI.

- User-to-user signaling procedures: A VCC may be set up within an existing VPC between two user network interfaces using user-to-user signaling. The intent of user-to-user signaling is to allow the two end users to inform each other of the nature of the applications running on each machine. An example of using this signaling procedure is VP tunneling, which will be introduced in Chapter 11.

During connection establishment, the user negotiates a QoS with the network and the remote end to set up appropriate traffic parameters. The network monitors the traffic to ensure that this agreement is followed.

3.3.3 ATM Cell Header Format

The ATM cell is the basic unit of information transfer in the B-ISDN ATM communication. The cell is comprised of 53 bytes. Five of the bytes make up the header field, and the remaining 48 bytes form the user information field. The ATM cell header consists of the following fields (depicted in Figure 3.9):

- GFC: The GFC field (four bits) is used to support multiplexing functions to alleviate short-term overload conditions when passing ATM traffic through a UNI interface. An NNI interface does not use this field for GFC purposes; rather, an NNI uses this field to define a larger VPI value for trunking purposes.

- VPI: Eight bits at UNI/12 bits for NNI.

- VCI: 16 bits.

- *Payload-type identifier* (PTI): Three bits. (A detailed description is given in Table 3.2.)

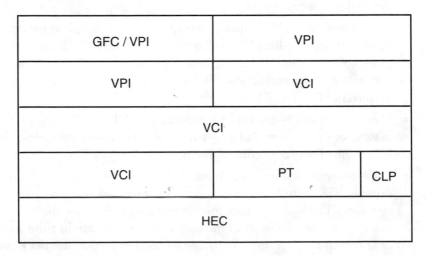

Figure 3.9 ATM cell header format. (*Source: ITU-T Recommendations, I.361, Figure 2.*)

Table 3.2

Payload Field Description

Type	PTI Coding 432	Description
User data cell	000	No congestion experienced, *service data unit* (SDU) type 0 (segmentation data unit)
	001	No congestion experienced, SDU type 1 (AAL5 end of packet)
	010	Congestion experienced, SDU type 0 (segmentation data unit)
	011	Congestion experienced, SDU type 1 (AAL5 end of packet)
Control cell	100	OAM F5 segment associated cell*
	101	OAM F5 end-to-end associated cell*
	110	Resource management cell
	111	Reserved for future use

* See Chapters 4 and 13 for further details.

- *Cell loss priority* (CLP): One bit; cells with this bit set are eligible to discard prior to cells in which the CLP is not set.

- HEC: Eight bits; HEC, a cyclic redundancy check code for the cell header field, is used for sensing and correcting cell errors and in delineating the cell header.

It should be noted that an ATM switch might not use all the VCI/VPI bits assigned by the standard due to the limitation of the resources available. At the UNI, 24 bits are assigned for connection identification, whereas there are 28 bits at the NNI. The actual number of routing bits in the VPI and VCI fields used for routing is negotiated between the user and the network. This number is determined on the basis of lower requirement of the user or network. A detailed description will be provided in Chapters 5 and 6.

Certain cell header values are pre-assigned for a number of uses by ITU-T [3] and the ATM Forum as listed in Table 3.3.

Table 3.3
Pre-assigned Header Values

Cell Type	VPI	VCI	PTI	CLP
Unassigned	0	0	XXX	0
Cell stream idle	0	0	XXX	1
VC resource management	VPI	Z	110	A
Meta-signalling	VPI	1	0A0	0
General broadcast signaling	VPI	2	0AA	= 0
Segment OAM F4	VPI	3	0A0	A
End-to-end OAM F4	VPI	4	0A0	A
Point-to-point signaling	VPI	5	0AA	0
VP resource management	VPI	6	110	A
Segment OAM F5	VPI	Z	100	A
End-to-end OAM F5	VPI	Z	101	A
ILMI message	0	16	AAA	0
LANE control	0	17	XXX	X
Private network-network interface routing	0	18	XXX	X

Note:
X denotes any value;
A signifies that the bit may be 0 or 1 and is available for use by the appropriate ATM layer function; and
Z denotes any value other than 0, 3, 4, 6, or 7.

3.4 ATM Adaptation Layer

As mentioned earlier, the ATM layer is independent of services that are provided by high layers. Thus the AAL [4, 5] is introduced to enhance the service provided by the ATM layer to support functions required by the next higher layer. The functions performed in the AAL depend upon the higher-layer requirements. Therefore, AAL is service-specific.

ITU-T has classified four types of services for ATM to support—class A, class B, class C, and class D—in terms of the timing relation between source and destination, bit rate and connection mode. Figure 3.10 lists AAL services defined by the ITU-T [4]. Consequently, four types of AALs were originally proposed to deal with each service. Type 1 was designed to adapt the constant bit rate services such as circuit emulation for 64-Kbps voice or constant bit rate video, which require a strict timing relation between source and destination. Type 2, which has not yet been standardized, was developed for variable bit rate services such as video and audio that also require a strict timing relation. Type 3 was intended for connection-oriented data transfer and type 4 for connection-less data transfer. It soon turned out that there was no need to maintain type 3 and type 4 separately, as their functions are similar. Accordingly, AAL 3 and AAL 4 have been merged and are referred to as AAL type 3/4. Subsequently, a new type, type 5, was defined to simplify the AAL type 3/4.

3.4.1 AAL Type 1

The AAL type 1 is used to adapt constant bit rate services over an ATM network including synchronous and asynchronous circuit transport, constant bit rate video, voice-band signal transport, and high-quality audio signal transport.

AAL 1 is responsible for packetizing a stream of bits or octets of data arriving at constant bit rate at the source for ATM layer transmission and restoring the original stream at its destination. The incoming data stream may be synchronous—i.e., the clock is frequency-locked to the clock of an ATM network—or asynchronous—i.e., the clock is not frequency-locked. Examples of the latter include PDH transport over ATM, such as ITU-T G.702 signals at 1,544, 2,048, 6,312, 8,448, 32,064, 44,736, and 34,368 Kbps. Examples of

	Class A	Class B	Class C	Class D
Timing relation between source and destination	Required		Not required	
Bit rat	Constant		Variable	
Connection mode	Connection-oriented			Connectionless
AAL protocol	Type 1	Type 2	Type 3/4, Type 5	Type 3/4

Figure 3.10 Service classification for AAL. (*Source: ITU-T Recommendations, I.362, Figure 1.*)

the former are 64, 384, 1,536, and 1,920 Kbps signals as described in ITU-T I.231. In either case, clock information should be transmitted so that the information can be restored at the receiving end. In addition, *forward error control* (FEC) is provided in AAL 1 to protect data. The detailed functions of AAL type 1 are listed in Table 3.4.

Since AAL 1 is quite complicated, a detailed description is beyond the scope of this book.

3.4.2   AAL Type 3/4

The AAL type 3/4 specifies the transfer of connection-oriented class C and connectionless class D services over ATM networks. The structure of AAL 3/4 is shown in Figure 3.11, in which the CS sublayer is further divided into two sublayers, the *service-specific CS* (SSCS) and the *common part CS* (CPCS).

Table 3.4
AAL Type 1 Functions

Plane	Functions
User plane	Segmentation and reassembly;
	Handling of cell delay variation;
	Handling of cell payload assembly delay;
	Handling of lost and misinserted cells;
	Source clock frequency recovery at receiver;
	Recovery of the source data structure at the receiver;
	Monitoring of AAL header for bit errors;
	Handling of AAL header errors;
	Monitoring of user information field for bit errors and possible correction.
Management plane	Indications that may be passed from the user plane to the management plane:
	Errors in the transmission of user information;
	Lost or misinserted cells;
	Cells with errored AAL protocol control information;
	Loss of timing and synchronization;
	Buffer underflow and overflow.

AAL	SSCS	Service-specific convergence sublayer
	CPCS	Common part convergence sublayer
	SAR	Segmentation and reassemble

Figure 3.11 Structure of AAL..

AAL 3/4 provided both assured operations and nonassured operations. Every assured AAL-SDU is delivered with exactly the same data content as sent by the user. The assured service is provided by retransmission of missing or corrupted SSCS-PDUs. Flow control is required in this mode. In nonassured operations, lost or corrupted AAL-SDUs are not corrected. Rather, higher layers are responsible for error recovery. The support of flow control for nonassured operations is optional. The AAL 3/4 CPCS layer only provides nonassured operations. The assured operations may be performed at the SSCS sublayer, which is defined in other standards such as Q.2931 signaling.

The functions provided at the SSCS sublayer depend on the services requested. They generally include functions for error detection and recovery and may also include special functions such as transparent delivery.

Two modes of services are defined in AAL 3/4: message mode and streaming mode. Message mode transfers framed data such as LAPD or *switched multimegabit data service* (SMDS) frames. In fact, AAL3/4 stems from SMDS. In this mode, the AAL SDU is passed across the AAL interface in exactly one *AAL interface data unit* (AAL IDU). While in the streaming mode an AAL-SDU is transmitted in one or more AAL IDUs. A pipeline function is supported, by which the sending AAL entity initiates the transfer to the receiving AAL entity before it has the complete AAL-SDU available. This reduces the size of the buffer required to store the IDUs belonging to the same AAL SDU.

The functions for AAL 3/4 are listed in Table 3.5. The format for AAL3/4 CPCS PDU is depicted in Figure 3.12. The CPCS header consists of one-octet *common part indicator* (CPI), one-octet *beginning tag* (Btag) and two-octet *buffer allocation size* (BASize) field. The CPI is used to interpret subsequent fields for the CPCS functions in the CPCS-PDU header and trailer. The Btag field allows the association of the CPCS-PDU header and trailer. The BASize field indicates to the receiving peer entity the maximum buffering requirements to receive the CPCS-SDU.

Table 3.5
Functions for AAL 3/4

Sublayer	Functions
CPCS	Preservation of CPCS SDU;
	Error detection and handling;
	Buffer allocation size;
	Abort a partially transmitted CPCS-SDU.
SAR	Preservation of SAR-PDU by making use of the segment type and payload length;
	Error detection and handling of bit errors, lost and misinserted SAR-PDUs;
	SAR-SDU sequence integrity;
	Multiplexing/demultiplexing;
	Paddling;
	Abort a partially transmitted SAR-PDU.

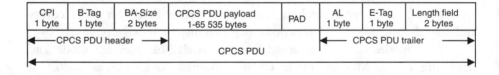

Figure 3.12 CPCS PDU format for AAL 3/4.

The CPCS trailer contains a one-octet *alignment* (AL) field, a one-octet *end tag* (Etag), and a two-octet length field. The sole purpose of the AL field is to achieve 32-bit alignment in the CPCS-PDU trailer. The length field indicates the length of the CPCS payload.

The sender inserts the same value into both Btag and Etag fields for a given CPCS PDU and changes the value for each successive CPCS-PDU. This allows a receiver to verify the integrity of the message received by checking the value of the Btag in the header with the value of Etag in the CPCS trailer.

Before a CPCS-PDU is segmented into one or more SAR PDUs, a padding field consisting of up to three octets is appended at the end of the information field in order to keep the CPCS-PDU payload to an integral multiple of 4 octets.

The format for AAL 3/4 SAR PDU is depicted in Figure 3.13. The header of SAR PDU contains a two-bit *segment type* (ST) field, a four-bit *sequence number* (SN) field, and a 10-bit *multiplexing indication* (MID) field. The ST indicates the type of the SAR in relation to a CPCS-PDU. There are four types of SAR PDUs. A *single sequence message* (SSM) contains an entire SAR SDU. If the SAR SDU is segmented into two or more SAR PDUs, the first one is the *beginning of message* (BOM), the last one is the *end of message* (EOM), and any intermediate SAR PDUs are termed *continuation of message* (COM). The SN allows the stream of SAR-PDUs of a CPCS-PDU to be numbered module 16. This can be used to detect cell misinsertion and cell loss. The MID field is used for multiplexing, which enables a number of CPCS connections to be multiplexed over one ATM connection.

The trailer of a SAR PDU consists of a six-bit *length indication* (LI) field and a 10-bit CRC field. The LI indicates the length of the SAR PDU information field. Another usage of this field is to abort a partially transmitted CPCS-SDU. A special value 63 is used to identify an abort-SAR PDU. The operation of AAL 3/4 is shown in Figure 3.14.

3.4.3 AAL type 5

AAL type 5 is also known as *simple and efficient AAL* (SEAL). As the name suggests, AAL 5 refines and simplifies the AAL 3/4 to reduce the protocol and transmission overhead and to ensure adaptability to existing transport protocols.

The structure of AAL 5 is the same as that of AAL 3/4, in which the CS sublayer is further divided into SSCS and CPCS sublayers. In addition, the services provided by AAL 5 are very similar to those offered by the AAL 3/4. The AAL 5 also supports both message mode and stream mode of services, as

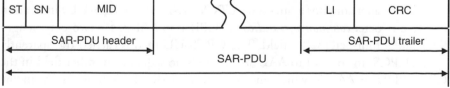

Figure 3.13   SAR PDU format for AAL 3/4. (*Source: ITU-T Recommendations, I.363, Figure 14.*)

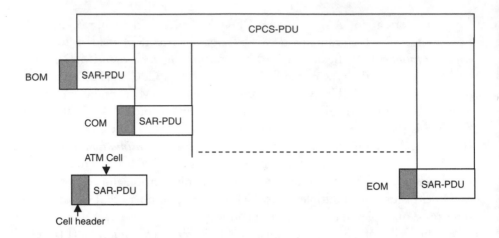

Figure 3.14 Transmission frame structure for AAL 3/4.

well as assured and nonassured operations. Again, the CPCS sublayer only performs nonassured operation.

Although AAL5 is classified as an AAL for connection-oriented services such as frame relay only, it is also widely used to support connectionless service. For example, AAL 5 has been adopted by the ATM Forum to support LANE, which is connectionless. Given the connection-oriented nature of ATM networks, connectionless services are actually provided by higher-layer protocols. As a result, there is nothing to prevent AAL 5 from being used to support connectionless services.

The functions for AAL 5 are listed in Table 3.6. The format for AAL 5 CPCS PDU is depicted in Figure 3.15. The CPCS PDU consists of the information field, which is up to 65,535 bytes, a padding (PAD) field, and an eight-octet trailer. The PAD, ranging from 0 to 47 octets, is appended to the information field so that the length of a CPCS PDU is an integral multiple of 48 octets.

The trailer contains a one-octet *CPCS user-to-user* (CPCS-UU) indication field, a one-octet *common part indicator* (CPI) as in AAL 3/4, a two-octet length field, and a four-octet CRC field. The CPCS-UU field is carried transparently by the CPCS. In contrast to AAL 3/4, there is no sequence number field in the CPCS-PDU in AAL 5, which means that all SAR PDUs must arrive in the proper order for reassembly. Since high-performance 32-bit CRC protection is provided at CPCS level, virtually all misinserted cells, cell loss, or bit error can be detected.

Table 3.6
CPCS Functions for AAL 5

Sublayer	Functions
CPCS	CPCS SDU delineation and transparent transmission;
	Transparent transmission of CPCS user-to-user information;
	Error detection and handling;
	Buffer allocation size;
	Aborting partially transmitted CPCS-SDUs;
	Padding;
	Handling of congestion information;
	Handling of loss priority information.
SAR	Preservation of SAR-PDU by making use of the segment type and payload length;
	Error detection and handling of bit errors, lost and misinserted SAR-PDUs;
	SAR-SDU sequence integrity;
	Multiplexing/demultiplexing;
	Aborting partially transmitted SAR-PDUs.

Unlike AAL 3/4, there is no overhead in the AAL 5 SAR sublayer. All 48 bytes of ATM layer cell payload are available for transmission of CPCS PDU. The PTI field of the ATM header is used to identify the types of SAR PDUs: 0 for BOM/COM and 1 for the EOM. The SAR PDU format for AAL 5 is shown in Figure 3.16. The operation of the AAL 5 is illustrated in Figure 3.17.

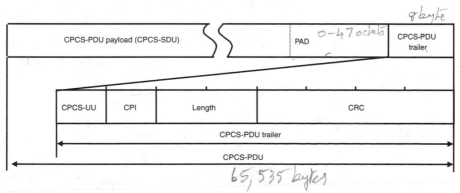

Figure 3.15 CPCS PDU format for AAL 5. (*Source: ITU-T Recommendations, I.363, Figure 6-5.*)

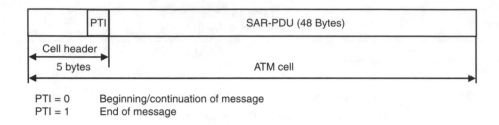

PTI = 0   Beginning/continuation of message
PTI = 1   End of message

Figure 3.16 SAR PDU format for AAL 5. (*Source: ITU-T Recommendations, I.363, Figure 6-4.*)

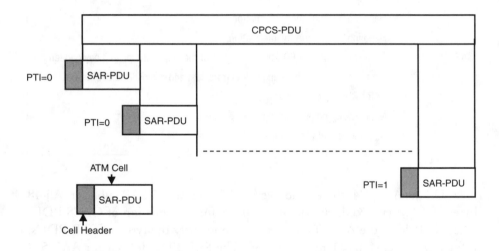

Figure 3.17 Transmission frame structure for AAL 5.

Compared with AAL 3/4, which has four bytes overhead for each cell, AAL only consumes eight bytes for each CPCS PDU. Thus, the overhead is minimized in AAL 5.

3.5 Addressing Model

An addressing model is essential for any network to identify its networked entities. The ATM addressing model offers a uniform addressing scheme for all ATM connected endpoints. There are two broad categories of addresses, public network addresses and private network addresses.

3.5.1 Public Network Addresses

Public network addresses are based on the ITU E.164 [6] ISDN numbering scheme. These numbers are usually assigned by the public telephony providers and are based on the geographical location of the user.

3.5.2 Private Network Addresses

E.164 addresses, a public thus expensive resource, cannot typically be used within private networks. The ATM Forum extended ATM addressing to private networks by defining an address format based on the syntax of an OSI *network SAP* (NSAP) [7] address defined in UNI specifications [8]. Note, however, that an ATM address is not an NSAP, despite the similar structure; while in common usage such addresses are often referred to as *NSAP addresses,* they are better described as *ATM private network addresses,* or *ATM end-point identifiers.* They identify not NSAPs but subnetwork points of attachment.

The ATM Forum's private network addresses (see Figure 3.18), also known as *ATM end system addresses* (AESAs), are fixed-length, 20-octet binary strings that are structured to reflect the network topology. The definitions for each field of AESA are as follows:

- *Authority and format identifier* (AFI): This is a one-octet field that specifies the format of the rest of the prefix and the authority responsible for allocating it. Three values are supported: international code designator (ICD), data country code (DCC), and E.164.

- *Initial domain identifier* (IDI): This specifies the authority responsible for allocating the rest of the prefix. Three types are defined based on the AFI as listed in Table 3.7.

- *High-order domain-specific part* (HO-DSP): This is the rest of the network prefix, and it may be structured hierarchically to reflect the network topology or address authorities. Subdomains may be created to facilitate address administration.

- End system identifier (ESI): A 6-octet field that identifies an end system and that is unique for a given network prefix. Generally, this is a 48-bit IEEE MAC address burnt into the ATM adapter card on the end system. The ESI along with the network prefix forms a unique 19-octet address in the network.

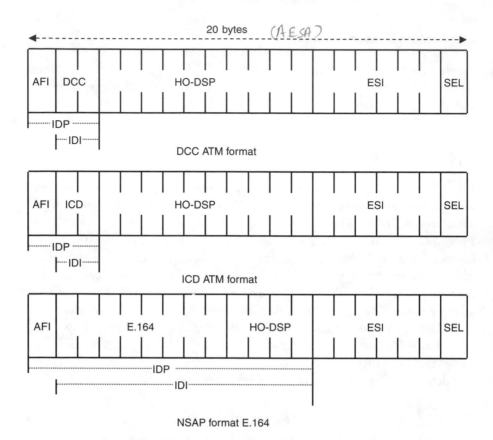

Figure 3.18 ATM private network address formats. (*Copyright 1996 The ATM Forum.*)

- *Selector:* A one-octet field that may be used by an end system to identify multiple user applications within the end system. It is not meant to
  be used within the network for any purpose (even, for example, call
  routing).

The following is an example of the NSAP address for an ATM switch:

0x47.00918100000000603E5ADB01.00603E5ADB01.00

The AFI field is 0x47, which means that the address is in ICD format
according to Table 3.7. The IDI field is 0x0091, specifying the authority
responsible for allocating the rest of the prefix, i.e., Cisco Systems Inc. The
example address is the default NSAP address generated by the Cisco Light-
Stream 1010 ATM switch during start-up. Its structure is shown in

Table 3.7
IDP Coding

Type	AFI (Hex)	IDI Length (Bytes)	Description
DCC	0x39	2	Specifies the country in which this address is registered. The codes are given in ISO 3166.
ICD	0x47	2	Specifies an international organization that is responsible for allocating the rest of the address. The British Standards Institute maintains the registration authority for ICDs.
E.164	0x45	8	Specifies an E.164 number, which is the same as that used in public networks. This may be used to specify a point of attachment of a private network with a public network, where the AESAs of hosts within the private network are derived from the E.164 address of the public UNI.

Figure 3.19. Note that the MAC address for the switch is 0x00603E5ADB01, which is used for both bytes 8 through 13 and bytes 14 through 19.

References

[1]   ITU-T Recommendation I.321, "B-ISDN Protocol Reference Model and Its Application," 04/91.

[2]   ITU-T Recommendation I.311, "B-ISDN General Network Aspects," 03/93.

[3]   ITU-T Recommendation I.361, " B-ISDN ATM Layer Specification," 03/93.

[4]   ITU-T Recommendation I.362, " B-ISDN ATM Adaptation Layer (AAL) Functional Description," 03/93.

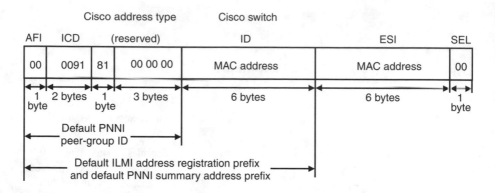

Figure 3.19 Cisco default ICD NSAP ATM address format.

[5]   ITU-T Recommendation I.363, " B-ISDN ATM Adaptation Layer (AAL) Specification," 11/93.

[6]   ITU-T Recommendation E.164, "Numbering Plan for the ISDN Era," 11/91.

[7]   ISO/IEC 8348, Information Technology, Telecommunications and Information Exchange Between Systems - Network Service Definition, 1993.

[8]   ATM Forum, ATM User-Network Interface Signaling Specification Version 4.0 af-sig-0061.000, July 1996.

4

ATM Network Management Architecture

4.1 Introduction

Network management functions are identified by the ISO in the management framework of the well-known OSI management model [1]. These functions can be classified into five areas as follows:

1. Configuration management: Configuration management involves determining what is in the network and manipulating those components to keep network interconnections and services up and running.

2. Fault management: Fault management encompasses detecting, isolating, and correcting abnormal network behavior. Due to the growing importance of the network in an enterprise, any interrupt of network services may affect the operation of the whole organization. Consequently, fault management is considered the most significant network management function.

3. Performance management: Performance management evaluates network behavior and effectiveness. Its aim is to understand how the network operates under normal and abnormal conditions. By monitoring the statistics of network components, the performance of a network can be fine-tuned, hence improved.

4. Security management: Security management deals with the monitoring, control, and protection of management information on the network from unauthorized or accidental access, disclosure, or modification. This functional area is also responsible for the management

of security facilities, including control and monitoring of them. Proper security management should log and report security threats or breaches.

5. Accounting management: Accounting consists of allocating costs among network users. This function includes collecting network usage information, setting accounting limits on network usage, and reporting on costs incurred.

A summary of network management functions for each category is given in Table 4.1. A detailed description of these functions can be found in [2]. The management functions are performed with the management plane of the B-ISDN protocol reference model. This book focuses on configuration, fault, and performance management. In Sections 4.2–4.4, the frameworks for ATM network management as well as ATM-specific management requirements will be introduced.

Table 4.1
Network Management Functions [3]

Category	Management Functions
Configuration management	Network status monitoring;
	Network routing;
	Parameter database;
	Configuration control;
	Facility control.
Performance management	Monitoring;
	Analysis;
	Database generation analysis;
	Reporting;
	Tuning.
Fault management	Event notification;
	Logging;
	Ticketing;
	Tracking;
	Isolation;
	Resolution.

Category	Management Functions
Security management	Authentication of users;
	Maintaining security;
	Encryption;
	Key distribution;
	Audits;
	Traces.
Accounting management	Issuing orders;
	Recording;
	Reconciliation of invoices;
	Development of cost algorithms;
	Assignment of costs.

4.2   ATM Network Management Reference Model

Figure 4.1 depicts the ATM network management reference model specified for end-to-end ATM network management by the ATM Forum's network management subworking group [4]. This model describes the various types of network management needed to support ATM devices, private networks, public networks, and the interaction between them.

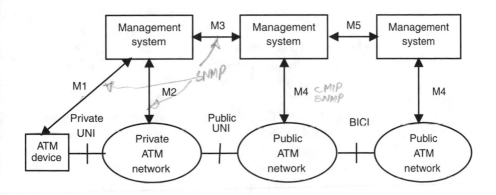

Figure 4.1 ATM network management reference model. (*Copyright 1996The ATM Forum.*)

The five key management interfaces defined in this framework—labeled M1 through M5—are all essential for end-to-end monitoring and control. They are described as follows.

- M1: The management interface between the private management system and the end ATM device (i.e., between an ATM *workstation* (WS) and the management system);

- M2: The management interface between the private management system and the switches making up the local private ATM network;

- M3: The management interface between the private management system and the public management system;

- M4: The management interface between the switches in the public ATM network and the public management system;

- M5: The management interface between two public management systems from different carriers.

M1, M2, and M3 embrace the Internet management framework, i.e., SNMP [4, 5], as it is the most widely deployed network management framework. SNMP is both the de facto and the de jure standard for today's private/ enterprise network management, although many other protocols are also useful, particularly Telnet and the *Internet control message protocol* (ICMP).

M1 and M2, which define the interface between the *network management system* (NMS) at the customer site and an ATM end-station or a private network switch, are of special interest for managers of enterprise networks. M1 and M2 interfaces encompass the relevant standard MIBs for the physical layer, ATM layer, and higher layers such as DS-1, DS-3, SONET/SDH, AToM MIB, and the MIB-II as defined by the IETF. The AToM MIBs and SONET/SDH will be described in Chapters 6 and 7, respectively. The AToM MIB allows network managers to group collections of switches, virtual connections, interfaces, and services into discrete entities and is SNMPv1- and SNMPv2-compliant.

Also encompassing these MIBs is the M3, the *customer network management* (CNM) interface. M3 describes the interface between the customer and carrier management systems that gives the customer a view into the carrier's network. M3 allows customers to supervise the use of their portion of a public ATM network. Ultimately, carriers plan to extend their CNM offerings so that the network managers can have real-time control over the services they use.

CNM for ATM public network service has been specified by the ATM Forum and will be addressed in Section 4.3.

Today, most LANs and the applications that run over them are SNMP-manageable. By making use of SNMP for M1 to M3, the management of private/enterprise networks is unified on SNMP framework, allowing existing management systems to work seamlessly with ATM NMSs, facilitating an integrated management for network, systems, and applications.

M4 is the management interface needed to manage a public network service, including both *network element* (NE) management and service management functions. This interface enables network and network element views from a carrier's NMS into the public ATM network it manages. Unlike private networks, public networks are committed to supporting the TMN where the *common management information protocol* (CMIP) has been adopted as the standard for transferring management information. As such, M4 is specified in such a way that different approaches can be used, including SNMP, CMIP, and possibly other emerging schemes. For each view, a logical MIB that specifies the management interface requirements is developed first. The purpose of defining a logical MIB is to provide a common frame of reference for the development of protocol-specific MIBs such as those based on CMIP or SNMP. The definition of protocol-specific MIBs from a common logical MIB should facilitate their potential coexistence within a public carrier's network. The ATM Forum has already developed the CMIP MIBs for both views.

M5 is the most complicated interface in the reference model since it is the management interface between NMSs from different carriers. There is no standard available for this interface yet.

Though the M2, M3, M4, and M5 management interfaces provide a top-down network view, they are not the only management functions relevant to ATM. The ILMI provides an ATM link-specific view of the configuration and fault parameters of an ATM interface. In addition, an ATM interface provides some layer management functionality by way of the OAM cells, which are detailed in Section 4.4. Customers can consider ILMI to be a CNM service since it allows information retrieval from a public network. However, the ILMI is embedded in the ATM interface and must be accessed by equipment directly connected to the interface. In contrast, the M3 interface allows direct access by customer premises-based management systems to CNM devices. ILMI will be introduced in Chapter 5.

There are also a number of SNMP MIBs that are defined to manage higher-layer functions or services running over ATM networks. These MIBs include LANE MIB, private network-network interface MIB, *circuit emulation service* (CES) MIB, and *data exchange interface* (DXI) MIB. Chapters 8 and 9 will be devoted to LANE MIB and PNNI MIB, respectively.

ATM remote monitoring (RMON) further enhances the manageability of ATM networks by adding high-level SNMP traffic monitoring capability to ATM systems. Based on RMON-standard MIBs, the ATM RMON MIB extends traditional RMON data sets such as flow statistics, host, and traffic matrix to ATM networks. This MIB allows a network manager to monitor network traffic for fault and performance monitoring and capacity planning.

4.3 Customer Network Management

CNM is a new service offered to customers by a telecommunications service provider (i.e., a career) that allows customers to access management information and functions existing within the provider's domain that relates to telecommunications services provided to the customer. CNM provides near real-time information about telecommunications parts of the network, enabling the private NMS to build and maintain a coherent, end-to-end view of the private network, its services, and performance. The functions in a typical CNM scenario [6, 8] are listed in Table 4.2.

Table 4.2
Typical CNM Functions (*Copyright CiTR Pty Ltd.*)

Category	Functions
Fault management	Reporting, tracking and resolution;
	Interface to customer trouble ticket or workflow system;
	Fault domain identification.
Configuration management	View inventory of *customer premises equipment* (CPE) and services provided by telecommunications provider;
	Order new services;
	Reconfigure services and network.
Accounting management	Expenditure tracking on services in near real time;
	Interface to customer accounting system;
	Extract of histories and usage profile by customer cost center, budgets, and authorization;
	Cost comparison of rival telecommunications services.

Category	Functions
Performance management	Monitoring of QoS parameters such as throughput, delay, and availability;
	Generating reports and verifying them against service contract;
	Performance comparison of rival telecommunications services.
Security management	Access authentication and authorization;
	Separation of carrier's and customer's data.

Given the diversity of services that a single customer may use and the variety of systems at the customer premises, a standardized interface is necessary for the exchange of management information. Figure 4.2 shows the architecture of CNM, in which the public network provider offers CNM service via a CNM agent in the provider's network. The CNM agent is, in turn, associated with an MIB that provides the required management information to the customer.

The carrier's network is monitored and controlled via its own NMSs. Part of this network is leased to customers to form part of their core enterprise network. Service-related information is passed from the carrier's NMS to the carrier's *service management system* (SMS). Each of the management entities within the carrier's system interacts with an agent adaptation function to update status in the carrier's CNM agent MIB. These systems include the carrier's accounting, NMS, SMS, or any other relevant carrier's legacy system. The function of the agent adaptation is to extract data from the different carrier sources. This is merged in the carrier's CNM MIB. The enterprise CNM manager receives alarms from the carrier's CNM agent, and the manager applications provide the necessary functionality to interface to the CNM agent MIB. The *business management system* (BMS) is involved in online reconfiguration and provisioning of new services.

There are some fundamental differences between traditional network management and the CNM [8]. First, the customer's NMS has only limited access to the information provided by the service provider. The service provider is responsible for managing the entire shared network as a whole, while service customers only view and manage their individual portions of the shared service. Because they have a restricted view of the network, customers are unable to

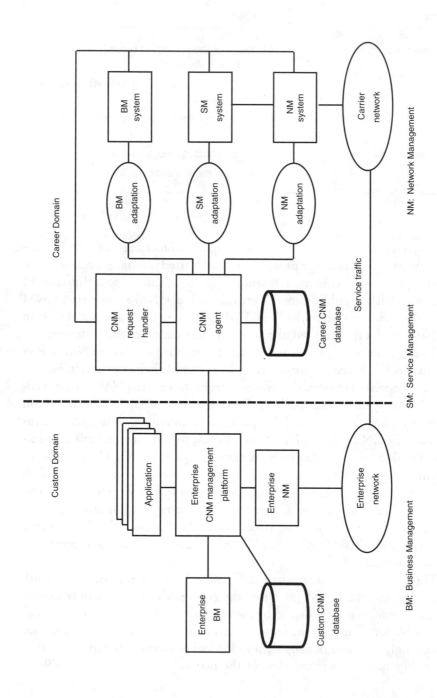

Figure 4.2 CNM architecture. (*Copyright CiTR Pty Ltd.*)

BM: Business Management   SM: Service Management   NM: Network Management

perform certain network management functions in the shared environment. For example, a customer that sets routes for optimized throughput of its own traffic may disrupt another customer's traffic. Only the service provider, with a complete view of the entire network, is in a position to determine routes that allow provisioned access to network resources for all customers.

Another fundamental difference in management functionality is that service providers manage the network internals directly, while customers manage their portion of the shared network indirectly. The service provider is responsible for the overall operation of the shared network, so any management control offered to customers must first be approved by the service provider before the control request takes effect in the network.

Finally, while service providers see a physical view of the network, customers see a logical view. This logical view includes the customer's configuration of service access points (i.e., logical ports) and the virtual connections that run between these logical ports. The customer does not see the individual network switches along the paths of its virtual connections, as the establishment of physical routes is a responsibility of the service provider.

The ATM Forum's M3 interface specification, "Customer Network Management for ATM Public Network Service" [4], defines CNM requirements, interface specification, and object definition for ATM permanent virtual connection service. The M3 specification only deals with network management; neither business nor service management functions have been defined. The agent's MIB models an ATM network as a large distributed switch by hiding all the network's internal connectivity as being internal to the distributed switch. Consequently, each concatenated virtual link that makes up a part of a virtual connection through an ATM public network is modeled as a virtual segment. With the M3 interface a customer is provided a view of a virtual ATM network that shows the customer's logical portion of the network.

Taking into consideration the differences between traditional network management and CNM, requirements for the M3 interface are classified into two categories to allow public network providers to offer modular incremental capabilities to meet different levels of customer need. Class I functions only provide read-only monitoring information, while Class II functions offer customers the ability to add, modify, or delete virtual connections and subscription information in a public ATM network.

The management framework to be used for M3 interface is the SNMP. The M3 interface specifies the use of MIB modules as shown in Table 4.3. All the modules are defined by the IETF; no new MIBs are defined by the forum for CNM.

Table 4.3
MIB Modules used for the M3 Interface

Layer	RFC No.	Abbreviation	Name
Transport and network	RFC 1213	MIB-II	Management Information Base for Network Management of TCP/IP-Based Internets: MIB-II
	RFC 1573	MIB-II Evolution	Evolution of the Interfaces Group of MIB-II
AAL ATM	RFC 1695	AToM	Definitions of Managed Objects for ATM Management, Version 8.0 Using SMIv2
Physical layer	RFC 1595	SONET/SDH	Definitions of Managed Objects for the SONET/SDH Interface Type
	RFC 1406	DS3/E3	Definitions of Managed Objects for the DS3/E3 Interface Type
	RFC 1407	DS1/E1	Definitions of Managed Objects for the DS1 and E1 Interface Types

4.4 Operation and Maintenance Functions

OAM is an integral part of a communications network. ATM provides an unprecedented level of OAM support by offering OAM functions at every level of the physical layer and ATM layer. ITU-T I.610 [9] specifies OAM principles and functions for B-ISDN.

In general, OAM functions deal with the following aspects of a network:

- Performance monitoring: Monitoring the normal operation of the managed entity by continuous or periodic checking of functions and collection of performance statistics;

- Defect and failure detection: Detecting malfunctions or predicted malfunctions by continuously, or periodically, generating various alarms;

- System protection: Excluding a failed entity from operation by blocking or changing over to other entities, so that the effect of failure of a managed entity is minimized;

- Failure or performance information: Transferring failure information to other management entities automatically or upon request;

- Fault localization: Identifying a faulty managed entity by an internal or external test system, if failure information is insufficient.

OAM functions in an ATM network are focused on performance monitoring, defect/failure detection, and localization, which can be categorized into performance and fault management. These functions are performed on five OAM hierarchical levels associated with the ATM and physical layer of the protocol reference model. The resulting bi-directional information flows F1, F2, F3, F4, and F5 are referred to as OAM flows [9], as shown in Figure 4.3. Note that not all of these flows need to be present. The OAM functions of a missing level are performed at the next higher level by the layer management entity. This does not mean that all OAM functions can be done at the highest level (consider, for example, F5). The purpose of including OAM flow at each level is to facilitate fast identification/recovery of fault conditions at each level. For example, once a fault condition is detected by physical layer OAM flows, the physical layer system will switch the service to a spare protection line so that the service is not interrupted. ATM layer OAM flows, which are shown in Table 4.4, are performed by special ATM cells known as OAM cells. The F4 and F5 flows give ATM network devices the ability to gather information about end-to-end connections.

4.4.1 Fault Management

Two types of ATM OAM cells, *alarm indication signal* (AIS) cells and *far end reporting failure* (FERF) cells, have been specified for fault management, which communicates failure information throughout the network. The operation of AIS and FERF cells is illustrated in Figure 4.4.

Whenever an ATM switch fails and a VP or VC is interrupted, each adjacent switch in the network automatically generates an AIS cell and sends it to all downstream switches. The AIS cell alerts the other switches in the network of the failure and gives them the opportunity to devise alternate routes for virtual connections that would normally cross the failed switch. FERF cells are

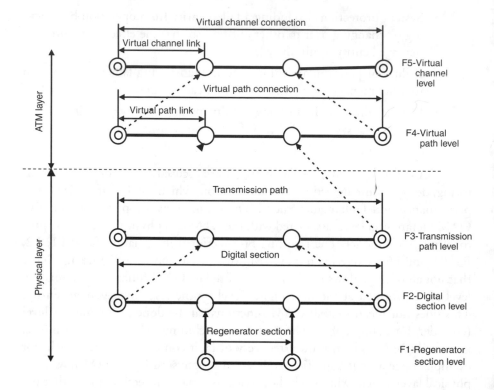

Figure 4.3 OAM hierarchical levels. (*Source: ITU-T Recommendations, I.610, Figure 2.*)

sent to the far end from a virtual connection endpoint as soon as it has received VP-AIS cells or detected connection failure.

The ATM Forum has also defined a fault management cell for continuity checking. When a VPC/VCC is idle for a certain period of time, end-stations or switches involved in the connection can send a continuity check cell to verify that the connection is still up.

In addition, an OAM loop-back capability, which uses a special loop-back cell, has also been specified for verifying connectivity and diagnostic problems that AIS or FERF cells cannot. The loop-back cell supports four loop-back scenarios as depicted in Figure 4.5, providing preservice connectivity verification, VPC/VCC fault sectionalization, and on-demand delay measurements. The loop-back capability is nonintrusive, which means that the loop-back test can be done online, without having to take the virtual connection out of service.

Table 4.4
OAM Functions of the ATM Layer

Level	Function	Flow	Defect/Failure Detection	System Protection and Failure Information
VP	Monitoring of path availability	F4	Path not available	For further study
			Degraded performance	
	Performance monitoring			
VC	Monitoring of channel availability	F5	Channel not available	For further study
	Performance monitoring		Degraded performance	

The loop-back cell is used in fault situations where an AIS or FERF cell would not indicate a problem, such as when a VPC/VCC has been misconfigured. In this case, there would be no actual failure in the network, and traffic would get through to both source and destination, but the connection would not be between the right end points. The loop-back cell, however, goes from source to destination and requires that the destination switch or end-station marks the cell and returns it.

4.4.2 Performance Management

Performance management of a VPC/VCC segment is performed by periodically inserting monitoring cells at the ends of the VPC or VPC segment. The inserted cells are capable of detecting such problems as errored blocks and loss/misinsertion of cells within a monitored block of cells.

4.5 Telecommunications Management Network

In addition to the Internet management framework, another important framework to manage an ATM network is the TMN. This framework was developed by the ITU to support management and deployment of dynamic

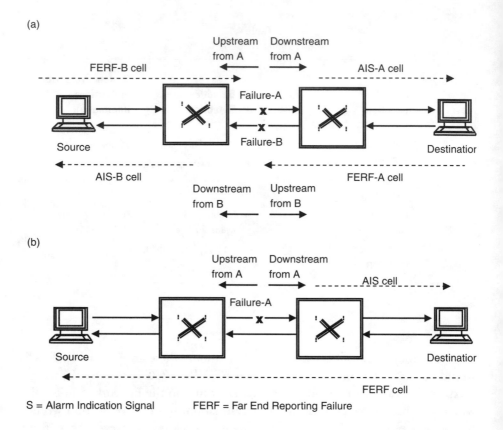

Figure 4.4   Operation of AIS and FERF cells: (a) failure in both directions and (b) failure in one direction.

telecommunications services. The TMN provides a host of management functions and communications for *operation, administration, maintenance, and provisioning* (OAM&P) of a telecommunications network and its services in multivendor environments.

Since ATM is the technology that is capable of implementing both LAN and WAN, there has been a lot of debate on which framework is better for ATM management. Traditionally, SNMP is the choice of LAN management, while TMN is the framework chosen by Telecoms for managing WANs. This section will introduce background for TMN and briefly compare the SNMP and TMN frameworks.

A telecommunications network is comprised of network resources such as switching systems, transmission systems, and terminals. In TMN terminology,

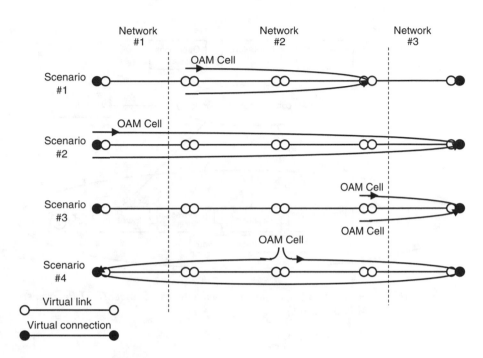

Figure 4.5 Loop-back scenarios. (*Source: ITU-T Recommendations, I.610, Figure I.1.*)

these resources are referred to as NEs. TMN enables communication between *operations systems* (OSs) and NEs via a *data communications network* (DCN) as shown in Figure 4.6.

4.5.1 TMN Architecture

The architecture of TMN [10] is defined in terms of three perspectives: a physical architecture, a functional architecture, and an information architecture. Sections 4.5.1.1–4.5.1.3 examine these perspectives.

4.5.1.1 TMN Physical Architecture and Interfaces

The physical architecture, which is depicted in Figure 4.7, consists of the following components:

- OS: The OS performs the management of NEs and services. It processes management information for monitoring, coordinating, and controlling telecommunications management functions. The OS can

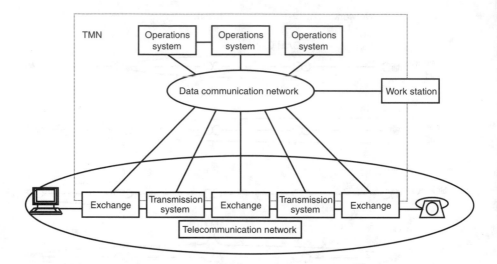

Figure 4.6 TMN in relation to a telecommunications network. (*Source: ITU-T Recommendations, M.3010 Figure 1.*)

also provide some of the mediation, Q-adaptation, and WS functions. NMSs, and *element management systems* (EMSs) are typical OSs.

- NE: The NE consists of the network components that are being managed. It contains manageable information that is monitored and controlled by an OS. To be managed within the scope of TMN, a NE must have a standard TMN interface. If a NE does not have a standard interface, it can still be managed via a *Q-adapter* (QA). The NE provides the OS with a representation of its manageable information and functionality (i.e., the MIB). As a building block, an actual NE can also contain its own OS function, QA function, *mediation device* (MD) function, etc.

- MD: The MD performs mediation between local TMN interfaces and the OS information model. A mediation function may be needed to ensure that the information, scope, and functionality are presented in the exact way that the OS expects. Mediation functions can be implemented across hierarchies of cascaded MDs.

- QA: The QA enables the TMN to manage NEs that do not have any TMN-compliant interfaces. The standard TMN interface is the

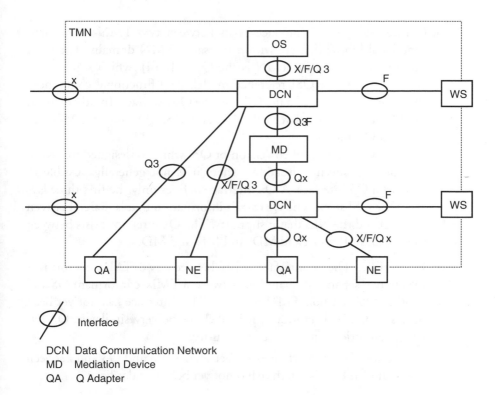

Figure 4.7 TMN components and interfaces. (*Source: ITU-T Recommendations, M.3010 Figure II-1.*)

common management information service (CMIS)/(CMIP) as will be discussed later in this section. The QA translates between TMN and non-TMN interfaces. For instance, an SNMP QA translates between SNMP and CMIP.

- WS: The WS performs WS functions (i.e., translating information between a TMN format and a displayable format for the user).

It should be noted that one physical TMN block may represent more than one TMN functional block. For example, a switch with a built-in management system also includes the OS function.

The information exchange between different TMN components are performed via standardized interfaces as listed below:

- Q-Interface: The Q interface exists between two TMN-conformant functional blocks that are within the same TMN domain. The most important interface for TMN is the Q3 [11, 14], which supports all seven layers of the OSI reference model. Any functional component that interfaces directly to the OS uses the Q3 interface. In other words, the Q3 interface is between the NE and OS; QA and OS; MD and OS; and OS and OS.

  The Qx is a simplified version of Q3, which is designed for use in environment-specific situations where it is not generally possible to support a Q3 interface. A Qx interface defines only the first three layers of the OSI stack. The Qx carries information that is shared between the MD and the NEs that it supports. The Qx interface exists between the NE and MD; QA and MD; and MD and MD.

- X interface: The X interface exists between two TMN-conformant OSs in two separate domains, or between a TMN-conformant OS and another OS in a non-TMN network. This interface has not yet been fully standardized. However, it is likely to be very similar to Q3 but with more strict requirements on security.

- F interface: The F interface exists between a WS and OS and between a WS and MD. This interface has not yet been standardized.

The relationship of the TMN components and interfaces between them are depicted in Figure 4.7.

4.5.1.2 TMN Functional Architecture

The functional architecture of the TMN introduces layering of TMN management functionality as follows:

- *Business management layer* (BML): The BML deals with such functions as high-level planning, budgeting, goal setting, executive decisions, and business level agreements.

- *Service management layer* (SML): The SML uses information presented by the NML to manage contracted service to existing and potential customers. This is the basic point of contact with customers for provisioning, accounts, QoS, and fault management. The SML is also the key point for interaction with service providers and with other administrative domains. It maintains statistical data to support QoS, etc. OSs in the SML communicate with OSs in the SML of other administrative domains via the X interface. OSs in the SML interface with OSs in the BML via the Q3 interface.

- Network management layer (NML): The NML has visibility of the entire network, based on the NE information presented by the EML OSs. In other words, the NML has the first *managed view* of the network. The NML coordinates all network activities and supports the demands of the SML. OSs in the NML interface with OSs in the SML via the Q3 interface.

- Element management layer (EML): The EML manages each NE. The EML has element managers, or OSs, each of which are responsible for the TMN-manageable information in certain NEs. In general, an element manager is responsible for a subset of the NEs. An element manager manages such areas as NE data, logs, and activity. Logically, MDs are in the EML, even when they are physically located in some other logical layer, such as the NML or SML. An MD communicates with an EML OS via the Q3 interface. In addition, an EML OS presents its management information from a subset of the NEs to an OS in the NML through the Q3 interface.

- NE layer (NEL): The NEL presents the TMN-manageable information in an individual NE. Both the NE and the Q-adapter, which adapts between TMN and non-TMN information, are located in this layer.

TMN management functions are categorized into functional blocks such as the *NE function* (NEF), *operation system function* (OSF), *mediation function* (MF), *WS function* (WSF) and *data communication function* (DCF). The same type of functions can be implemented at many levels, from the highest level, which manages a corporate or enterprise network, to a lower level that is defined by a network or network resource. Once management is defined at the lower layers, additional management applications can be built on this foundation.

A TMN physical building block contains one or more functional blocks that may communicate among themselves within the same building block or with function blocks in other building blocks. When function blocks in different building blocks communicate with one another, a TMN interface is used.

The key function in TMN is the OSF, which is responsible for such purposes as monitoring, coordinating, or controlling telecommunications

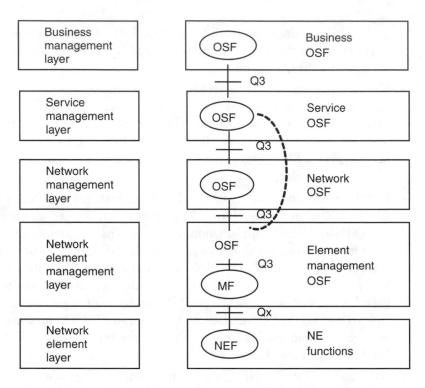

Figure 4.8 Hierarchy of the OS function. (*Source: ITU-T Recommendations, M.3010 Figure III-1.*)

functions. The TMN OS functional hierarchy is illustrated in Figure 4.8. Note that in some instances it may be possible for administrations to bypass layers of communication within the functional hierarchy—in order to reduce the response time, for example.

4.5.1.3 TMN Information Architecture

The TMN information architecture is based on the existing OSI management framework. The core standards include the following.

- CMIP [15]: defines management services exchanged between peer entities;

- *Guideline for definition of managed objects* (GDMO) [16]: Provides templates for classifying and describing managed resources;

- ASN.1: Provides syntax rules for data types.

The key to the OSI management framework is the use of the *object-oriented* (OO) approach, with managed information in network resources modeled as attributes in managed object classes. The main properties of an object class include operations permitted, notifications emitted, attributes data contained, and behaviors executed.

Object-orientation is the technique that is generally considered to be superior in dealing with the complexity of network management tasks. The OO approach is typically characterized by features such as encapsulation, inheritance, object class, and allomorphs. Management information is encapsulated in managed objects, thereby hiding internal implementations from a user. The object class extends and combines object definitions to create multiple instances of an object class. Objects are organized into a tree-like structure via containment relationships. Object definitions can be extended by inheritance, which allows a new object class to be derived from an existing object class through extension, modification, or restriction, permitting the reuse of object definitions. In particular, a new object can acquire attributes, notifications, operations, and behavior from objects higher in the tree. Allomorphs allow an entity to take many forms, providing a mechanism for migration and coexistence between multiple versions of class and object definitions. Although managed resources in SNMP are referred to as objects, the information model in SNMP is not OO. This is because the model does not support the typical characteristics mentioned above. Please refer to [17] for further discussion.

4.5.2 TMN-Based ATM Management

TMN-based network management involves a generic information model that defines technology-independent TMN object classes [18] and an ATM-specific information model that defines ATM specific object classes. The management information is specified in terms of several different viewpoints as follows:

- NE View: Concerned with the information that is required to manage a NE including the information required to manage the NEF and the physical aspects of the NE;
- Network View: Concerned with the information representing the network, both physically and logically, including how NE entities are related, topographically interconnected, and configured to provide and maintain end-to-end connectivity;
- Service View: Concerned with how network view aspects (such as an end-to-end path) are utilized to provide a network service, including

the requirements of a network service (e.g. availability, and cost), how these requirements are met through the use of the network, and all related customer information.

ITU-T Recommendation I.751 [19] specifies the Q3 interface between an ATM cross-connect and the NMS. It provides management requirements and an information model that pertain to the plane management of the ATM NE. The information model describes the managed object classes and their properties that are used to describe the information exchanged across management interfaces defined in the ITU-T M.3010. The I.751 specializes the generic object classes that have already been defined in the TMN framework to provide the information model specific to the ATM NE.

The ATM Forum network management working group has also defined a set of standards for the management of ATM network based on TMN including the following:

- af-nm-0020.000: "M4 Interface Requirements and Logical MIB; ATM NE-view," specifying the interface requirements and a logical MIB for the M4 network element view interface;

- af-nm-0027.000: "CMIP Specification for the M4 Interface," specifying the Q3 protocol stack and CMIP MIB for the M4 network element view interface;

- af-nm-0071.000: "AAL Management for the M4 'NE View' Interface," specifying a set of requirements, logical information model, and CMIP specification to support AAL management;

- af-nm-0072.000: "CES Interworking M4 Interface 'NE View' Requirements, Logical and CMIP MIB," specifying the requirements, protocol-independent managed entities, and CMIP MIB specification for circuit emulation service for the M4 network element view interface;

- af-nm-0058.000: "M4 Network-View Interface Requirements, and Logical MIB," specifying the improvements to the M4 network-level protocol-independent MIB (AF-NM-0058.000);

- af-nm-0074.000: " M4 Network View Requirements & Logical MIB Addendum," specifying the interface requirements and a logical MIB for the M4 network view interface;

- af-nm-0073.000: "M4 Network View CMIP MIB Specification," specifying the CMIP MIB for the M4 network element view interface;

ATM-Forum M4 and M5 interfaces are modeled as TMN interfaces. The M4 interface is a TMN Q3 interface between a public NMS and a TMN NE, an EMS, or another NMS. The M5 interface between two public NMSs of different network providers is a TMN X interface, i.e., an interface between OSs in different TMNs.

4.5.3 CMIP Versus SNMP

Both Internet SNMP and OSI CMIP provide an open, standard platform to perform interoperable network management in a multivendor, heterogeneous environment. They are very similar in terms of their functionality and basic approach.

CMIP and SNMP both use a client-server model in which the client is the managing system and the server is the managed system. The managed system (server) assumes an agent role, receiving management requests, executing commands, and transmitting unsolicited event notifications/traps. The managing system (client), acting in the role of a manager, invokes operations and receives notifications/traps. ASN.1 is used in both frameworks to describe management information. However, there are some essential differences between the frameworks as described in Sections 4.5.3.1–4.5.3.5.

4.5.3.1 Information Model

The OSI management framework is based on the OO design principles in representing management information. The resources managed are presented in terms of managed objects and managed object classes. With OO concepts such as inheritance and allomorphs, management information can be organized in an orderly and structured manner, thus permitting a higher level of abstraction. This in turn allows the OSI information model to scale to large networks.

In contrast, MIB structure is relatively flat in SNMP. Management operations are restricted to scalar objects. The only constructed type supported is a simple two-dimensional conceptual table, consisting of only scalar objects. Thus it is difficult, if at all possible, to implement tables inside a table.

The object naming structure for SNMP is straightforward. Every object type has a registered OID, and an object instance is identified by the object type OID suffixed by a part that uniquely identifies that instance. By employing such a scheme, however, no generic object can be defined. For example, a RowStatus object has to be defined for each table that permits row creation. Unlike SNMP, CMIP does not have such a restriction; this allows a generic RowStatus-like object to be used in any table.

Finally, although managed resources are also referred to as objects, SNMP is not OO. For example, an SNMP object cannot inherit from another object. OO was considered by early SNMP developers as creating unnecessary complications without real benefits. SNMP objects are different from the objects that are defined in CMIP, in which the SNMP-equivalent managed information in network resources is modelled as the attributes of managed objects. CMIP notifications are defined as properties of the managed objects. In brief, the OSI information model is more flexible and powerful than that of SNMP.

4.5.3.2 Protocol

Both CMIP and SNMP are application layer protocols. CMIP is based on the well-known ISO-OSI reference model, which consists of seven layers. SNMP, on the other hand, conforms to the Internet four-layer protocol stack. The SNMP protocol stack is simpler than the OSI stack.

SNMP protocol messages are transferred over connectionless UDP. The prime motivation behind this choice is that SNMP must continue to operate (if at all possible) when the network is operating at its worst [20]. By using a connectionless transport protocol, SNMP takes on the responsibility of reliable data transmission. This means that an SNMP application may time out outstanding requests and either retransmit them or abort them as appropriate—this is normally done by a transport protocol. In contrast, CMIP makes use of a connection-oriented reliable service provided by the OSI transport layer. This service can be implemented over a variety of transport and network layer protocol combinations.

The protocol operations of SNMP have been deliberately specified to be very simple. SNMP-managed objects accept only Set and Get operations while imperative commands and table creation/deletion are emulated through Set. The CMIP provides a richer set of operations, including Get, Set, Action, Create, Delete, Event-report, and Cancel-get, permitting sophisticated access to managed objects. Imperative commands (e.g., Action, Create, and Delete) can be executed directly. Moreover, mechanisms such as scoping and filtering greatly reduce network management traffic. With scope, for example, CMIP can issue a single request to retrieve all the information needed, avoiding the problem of multiple requests that SNMP requires. By making use of filtering capabilities, threshold values in an agent can be set to eliminate the transmission of routine information to a management station. In contrast, SNMP has no such mechanisms. As a result, SNMP requires a series of requests from the management station, transmitting normal as well as abnormal data. The filtering of unwanted data is the responsibility of the network management station. It should be noted that scoping and filtering has a performance penalty on the agent, because of the complexity. Consequently, this mechanism has not been widely implemented. Despite the complicated mechanism used, the actual efficiency of moderately scoped/filtered CMIP protocol is not so high as expected as a result of the overhead of the protocol stack.

4.5.3.3 Management Philosophy

Both SNMP and CMIP accomplish the role of management in the overall distributed architecture, albeit in different ways. The idea behind the ISO management framework is distributing intelligence over a network (e.g., among managed and managing entities). The resulting solution is thus more flexible but requires more resources in terms of memory and processing capability in agents. Consequently, the entry cost for the first application is relatively expensive, but additional objects can be added at sometimes relatively smaller incremental costs.

In SNMP, the design strategy is to shift workload toward management stations so that the impact of adding network management to managed devices is minimal, reflecting a lowest common denominator. As a result, an SNMP agent can be implemented even in the smallest/cheapest of systems. Any required complexity is performed by an SNMP manager; hence, the operational role and performance of a system is not compromised by the inclusion of an SNMP agent. Because the number of management stations is far less than that of management agents, overall system complexity and cost are optimized.

As the SNMP traps are sent asynchronously from managed devices to network management stations and are not retransmitted, they are inherently unreliable. Accordingly, network mangers should not rely on them. Alternatively, most implementations use polling to check the status of agents. Although *trap-directed polling* is recommended, some implementations only use polling in order to reduce the complexity of management applications. This limits the scalability of SNMP to very large networks due to the amount of processing and traffic generated for polling managed devices.

The CMIP event reporting mechanism is very sophisticated, allowing for very flexible control of emitted notifications. CMIP-based network management adopts an event-driven mechanism as it employs a reliable OSI transport service and as agents are more intelligent. Event reports are generated to report important spontaneous events. Event reports also supply performance information. In this way, network overhead is considerably reduced, thus extending the scalability of NMS. However, regular polling is necessary under all conditions to ensure the proper operation of a network.

4.5.3.4 Interoperability

The biggest strength of SNMP is its widespread popularity. SNMP-based network management has been accepted as the Internet standard management framework and has become deeply entrenched as the management framework of choice; it is now in pervasive and continuous use. Many MIB modules have been produced for a plethora of uses in network management, system management, application management, proxy management of legacy devices, and manager-to-manager communications. These MIBs reach virtually every kind of network technology. Now SNMP is taking over additional market segments while growing strong in the areas of telecommunications management, data and voice over cable, system management, and application management. As such, SNMP-based network management can be easily extended to applications that run over the managed network.

The application of CMIP, on the other hand, is largely limited within the context of TMN. Though more and more CMIP MIBs have been specified to manage a number of telecommunications networks, their scope is limited mainly to new networking technologies, such as ATM and SONET.

4.5.3.5 Suitability for ATM Management

The OO approach and the scalability of the OSI network management framework are distinctive advantages in managing large networks. Higher-level data is reflected automatically in the lower-level data because of inheritance and containment. For example, if an interface is down, all the protocols running over the interface are down. The scoping and filtering capability is especially useful when used in high-speed ATM networks, as a lot of alarm conditions may occur due to a link failure. This framework is made more attractive by the introduction of TMN with CMIP specified as the management protocol, as most carriers are committed to TMN.

However, the cost for the implementation of CMIP agents is much higher than that for SNMP, as the former is much more complex and requires many more resources than the former. As a result, the performance penalty on the agent (e.g., reduced throughput) may be significant. This effectively prevents CMIP from being used in other areas, such as data communications equipment. SNMP, on the other hand, offers a low entry cost, simplicity in implementation, and a fast standardization process.

Admittedly, SNMP is by no means perfect. SNMP was not designed to handle the huge volumes of network traffic that ATM generates. Some adaptation may be necessary for ATM management. For example, it may be necessary to use middle-level managers to correlate alarms generated by the switches in a large network to reduce management traffic. However, SNMP is well understood and widely used. A lot of SNMP MIBs have already been defined for ATM. These MIBs cover almost all aspects of ATM. Moreover, SNMP is well-balanced in terms of management functions and resources required, providing a cost-effective and interoperable solution to the challenges of managing ATM networks. Most importantly, it is a proven technology. Almost all ATM switches today are SNMP-manageable. Similarly, most ATM applications are, or will be, running over TCP/IP that is managed by SNMP. As a result, the scope of SNMP-based ATM management can easily be extended to non-ATM devices and applications running over ATM. In addition, SNMP is evolving, and some of its deficiencies, such as security, have been overcome in newer versions of the protocol. SNMP is an interoperable solution for ATM network management.

Taking into consideration the work that has been undertaken in the ATM Forum, CMIP and SNMP will coexist at least in the near future, although in different areas. SNMP will be used in access/private networks to allow existing NMSs and user applications to be integrated seamlessly, while CMIP will most likely be used in public telecommunications networks.

References

[1] ITU-T Recommendation X.700: "Management Framework for Open Systems Interconnection (OSI) for CCITT Applications," 09/92.

[2] ITU-T Recommendation M.3400: "TMN management functions," 10/92.

[3] Held, G., *Network Management, Techniques, Tools and Systems,* New York, NY: John Wiley & Sons, 1992.

[4] ATM Forum, Customer Network Management (CNM) for ATM Public Network Service (M3 Specification), October 1994.

[5] Peter Alexander and Kacey Carpenter, "ATM Net Management: A Status Report," Data Communications on the Web, September 1995, http://www.data.com/Tutorials/ATM_Net_Management.html.

[6] ITU-T Recommendation X.160, "Architecture for Customer Network Management Service for Public Data Networks, 10/96."

[7] ITU-T Recommendation X.161, "Definition of Customer Network Management Services for Public Data Networks," 04/95.

[8] Holliman, G., M. Hinchliffe, N. Cook, and P. Barnes, "Customer Network Management," *Data Communications International,* Vol. 24, No. 15, October 1995.

[9] ITU-T Recommendation I.610, "B-ISDN Operation and Maintenance Principles and Functions," 03/93.

[10] ITU-T Recommendation M.3010, "Principles for a Telecommunications Management Network," 10/92.

[11] ITU-T Recommendation Q.811: "Lower Layer Protocol Profiles for the Q3 Interface," 03/93.

[12] ITU-T Recommendation Q.812: "Upper Layer Protocol Profiles for the Q3 Interface," 03/93.

[13] ITU-T Recommendation Q.821: "Stage 2 and Stage 3 Description for the Q3 Interface—Alarm Surveillance," 03/93.

[14] ITU-T Recommendation Q.822: "Stage 1, Stage 2 and Stage 3 Description for the Q3 Interface—Performance Management," 04/94.

[15] CCITT Recommendation X.711, "Common Management Information Protocol (CMIP) Specification," 1991.

[16] ITU-T Recommendation X.722, "Information Technology—Open Systems Interconnection—Structure of Management Information: Guidelines for the Definition of Managed Objects," 09/92.

[17] Rose, M. T., *The Simple Book: An Introduction to Internet Management,* Second Edition, Englewood Cliffs, New Jersey: Prentice-Hall, 1994.

[18] ITU-T Recommendation M.3100, "Generic Network Information Model," 10/92.

[19] ITU-T Recommendation I.751: "Asynchronous Transfer Mode (ATM) Management of the Network Element View," 03/96.

[20] Kastenholz, F., Internet Engineering Task Force, RFC 1270, "SNMP Communications Services," October 1991.

5

Integrated Local Management Interface

5.1 Integrated Local Management Interface

The ILMI [1, 2], formerly known as the interim local management interface, was defined by the ATM Forum to ensure the interoperability of ATM switches from different vendors. When the ILMI was originally specified, it was expected to be used for only a short period of time—i.e., until the ITU-T management specification was completed—hence the name interim. However, the Forum now expects that the ILMI will be used "indefinitely."

The purpose of the ILMI is to enable two adjacent ATM devices to automatically configure the operating parameters of the common ATM link between them and to exchange management information. In particular, this interface is used to provide any ATM device with status and configuration information concerning the physical layer interface, ATM layer interface, and the VPCs and VCCs of the adjacent system. ILMI was initially defined for public UNI and private UNI. However, the latest version, ILMI 4.0 [2], has been extended to support private NNI. The ILMI fits into the overall management model for an ATM device as illustrated in Figure 5.1. In addition, the ILMI specification provides both users and network operators with network management functions that are not currently available from other technologies in a standard and consistent manner. These capabilities include bi-directional communication across the ATM link enabling both sides to verify subscription parameters, the configuration of new traffic services such as *available bit rate* (ABR), signaling version identification, and address registration information.

131

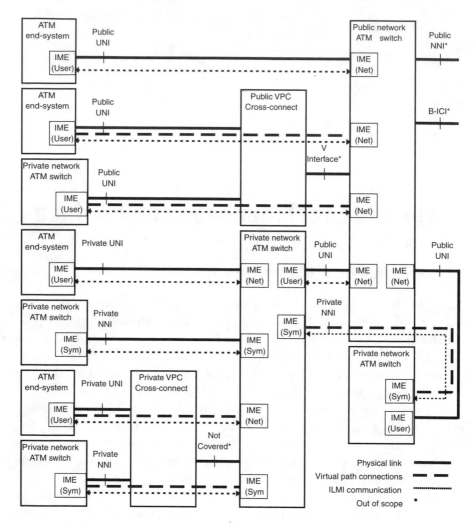

Figure 5.1 The ILMI. (*Copyright 1996The ATM Forum.*)

An *interface management entity* (IME) is associated with each ATM inter-face that supports the ILMI functions for that ATM interface. There is a per-ATM interface set of managed objects representing the ATM interface ILMI attributes that is sufficient to support the ILMI functions for each ATM inter-face. The ATM interface ILMI attributes are organized in a standard MIB structure; there is one MIB structure instance for each ATM interface on each ATM device. Consequently, an ATM device may have multiple IMEs if it sup-ports multiple ATM interfaces.

5.1.1 ILMI Protocol

ILMI communication takes place between adjacent IMEs over physical links or virtual links such as VPCs used by the *interim interswitch signaling protocol* (IISP) or PNNI. Rather than defining a new protocol, the ILMI is designed to use the same message formats and semantics as are used by SNMPv1. Thus, it has the same five PDU-types as SNMPv1 (get, get-next, set, get-response, and trap), and the variables included in these PDUs are defined in MIBs using SNMPv1's SMI.

The ILMI communication protocol is an open management protocol (i.e., SNMP/AAL) as shown in Figure 5.2. An IME can access, via the ILMI communication protocol, the IME ILMI MIB information associated with its adjacent IME. A well-known VCC is defined for the ILMI. The default value for provisioning this VCC is VPI = 0, VCI = 16, but the VPI/VCI value is configurable.

However, the ILMI has a paradigm different from the use of SNMP for network management. In network management, a manager issues requests and an agent responds with responses and generates traps. In contrast, the ILMI is used for interface management between the two IMEs, one on either side of the interface; both IMEs can send requests as well as responses and traps; the two IMEs have their own (potentially different) values for the same MIB objects.

ILMI 4.0 supports link configuration for UNI, IISP, and PNNI. In addition, the ILMI 4.0 specification provides new network management functions for both users and network operators that are not currently available from other standards. These capabilities include bi-directional communication across the ATM link enabling both sides to verify subscription parameters, the configuration of new traffic services such as ABR, signaling version identification, and

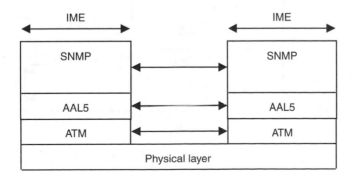

Figure 5.2 ILMI information transmission.

address registration information. A comparison between this MIB and RFC 1695 will be described in Chapter 6. ILMI also provides functions for LANE to auto-configure a *LANE client* (LEC). For further details on LEC auto-configuration, please refer to Chapter 9.

5.1.2 Proxy Agent

In addition to its role for local interface management, the data in ILMI MIBs is also useful for general network management functions. ILMI defines a proxy-agent mechanism that uses the existing functions of the ILMI to enable NMS access to the ATM interface MIB. The proxy uses the agent-role capability of the local IME to access ATM interface MIB data in the local system and the manager-role capability of the IME to access ATM interface MIB data in the neighboring system. The solution is depicted in Figure 5.3.

The SNMP proxy-agent accepts SNMP requests from an NMS and relays them to the appropriate IME for processing as either local operations to be run

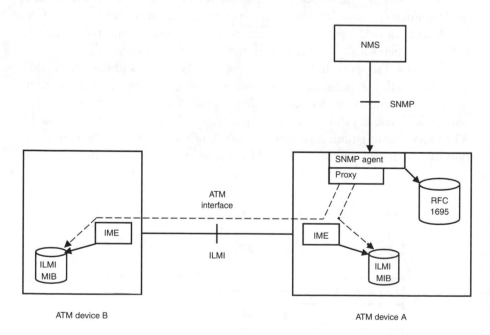

Figure 5.3 NMS access to ILMI MIBs through proxy-agent. (*Copyright 1996 The ATM Forum.*)

against its own ATM interface MIB or as operations to be forwarded across the ILMI interface to its peer IME for remote processing.

For each IME, two proxy-targets are defined, one to receive requests for the local ATM interface data and the other to handle requests for data in the neighboring ATM interface MIB. When the proxy-agent receives a request from an external NMS, it first determines in terms of the community-string in the request whether it should handle the request normally or whether the request should be forwarded to one of the IME proxy-targets. For example, a request with community string "community" identifies the message to be sent to RFC 1695 as usual. If the string is "local," the request will be forwarded to the local IME, while the "remote" packet will be sent to the adjacent IME.

For SNMPv1 requests, the proxy-agent does not modify the PDU it forwards between the NMS and proxy-target. SNMPv2 requests are forwarded to the IME using the SNMPv2-to-SNMPv1 proxy procedures defined in RFC 1908.

When the proxy-target receives the request, it performs the SNMP operation with respect to its defined MIB view and sends a response to the proxy-agent, which, in turn, sends it to the NMS.

5.2 ILMI MIB

ILMI MIB consists of four modules as follows:

1. Textual conventions MIB: defines a number of common textual conventions and OIDs in a single module so that other MIB modules may import them in a mutually consistent fashion.

2. Link management MIB: provides a general-purpose link management facility for ATM interfaces.

3. Address registration MIB: contains the information necessary for address registration.

4. Service registry MIB: extends the ATM interface MIB to provide a general-purpose service registry for locating ATM network services such as the *LANE configuration server* (LECS).

Section 5.2 focuses on the link management, address registration, and service registry MIBs.

5.2.1 Link Management MIB

5.2.1.1 MIB Structure

The link management MIB, which offers a general-purpose link management facility for ATM interfaces, provides the groups as shown in Figure 5.4. The MIB introduced in this chapter is based on ILMI 4.0. Since ILMI is an evolving standard, some objects that were introduced in its earlier versions are no longer used. However, ILMI 4.0 still inherits them for compatibility.

5.2.1.2 System Information

No per-system object is defined in the link management MIB. Instead, IMEs implementing the ATM layer interface group must support the system group defined in RFC 1213 MIB-II as described in Chapter 2. An IME must provide access to the system group via the ILMI communications protocol.

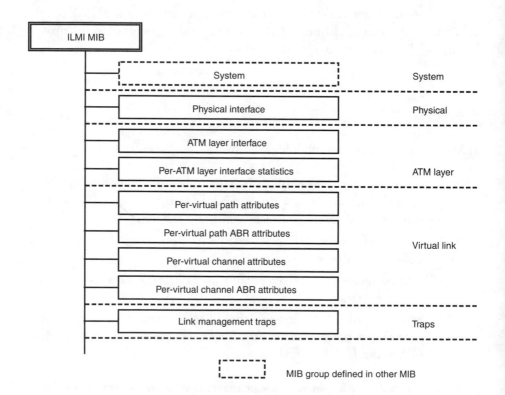

Figure 5.4 ILMI MIB structure.

5.2.1.3 Physical Interface

The ILMI provides access to management information identifying the physical layer interface. When ILMI communication takes place over a physical link, there is one physical layer group object for that physical interface. When ILMI communication takes place over a virtual link (i.e. a VPC used by PNNI), the physical layer management information is present and represents the virtual interface. This group contains a set of specific attributes and information associated with each physical or virtual interface.

The physical port group specified in versions 3.1 and early versions of the ILMI specifications provides configuration information and statistics about the ATM interface's physical layer interface. Most of these objects have now been deprecated. This information is available via standard network management MIBs—for example, SONET MIB if the underlying physical layer is SONET.

MIB information at this level includes the following:

- Interface index;
- Interface address;
- Operational status;
- Physical port information;
- Adjacency information.

This group consists of a set of tabular per-interface attributes and nontabular attributes, as listed in Tables 5.1 and 5.2, respectively. The tabular per-interface attributes are stored in the atmfPortTable with each entry per-interface, while the nontabular attributes are common to all the interfaces.

Each IME has a physical ATM interface or a virtual interface associated with it. The physical layer attributes of an ATM interface are stored in an entry in the atmfPortTable. The index object was designed to support the combination of all UNI ILMI MIB tree structures of a device with multiple UNIs into a single MIB, for use by a general NMS.

Note that the usage of atmfPortIndex is different from that of MIB II ifIndex. The normal range for ifIndex is (0..2147483647). In ILMI, however, a distinctive value zero is used to identify the physical or virtual interface over which ILMI messages are received. Moreover, only implicit identification is allowed. Explicit identification of interfaces using non-zero values is not allowed. Versions 3.1 and older of the UNI specification optionally allowed one of many interfaces to be explicitly identified by an interface index unique

Table 5.1
Per-Interface Objects for the Physical Layer Group

Prefix	atmfPort		
Category	Name	Definition	Comments
Index	Index	Interface index to identify a particular physical or virtual interface on the ATM device	"0" identifies the ATM interface over which SNMP messages are received
Operational status	OperStatus	The operational state of this port, including other, inService, outOfService, and loopBack.	
Interface address	Address	The interface address object specified in version 2.0 of the UNI specification	Obsolete; the address group defined in the address registration MIB should be used instead
Physical Port Info	TransmissionType	Type of physical transmission, such as SONET or DS3	Deprecated
	MediaType	Media type such as coax, single, or multimode fiber, STP, or UTP	
	Specific	A reference to additional information specific to the transmission type	
Adjacency information	MyIfName	A textual string uniquely names this interface	These two objects can be used to distinguish parallel links with a neighboring system
	MyIfIdentifier	A value uniquely identifies this interface	

to the ATM device. This option has been deprecated because it led to interoperability problems.

Table 5.2
Nontabular Objects for the Physical Layer Group

Prefix	atmfMy		
Category	Name	Definition	Comments
Adjacency information	IpNmAddress, OsiNmNsapAddress	An IP or NSAP address to communicate with a NMS	A typical IME supports only one of the these two objects
	SystemIdentifier	A 48-bit IEEE MAC address for this ATM device	ESI field in AESA

Topology Discovery

ILMI makes it possible for network topology discovery by making use of adjacency information defined in this group. The ATM auto-discovery functionality is important in determining the initial physical connectivity of the network and ongoing topological changes.

The adjacency information objects allow the neighboring system to maintain a table of adjacent systems to facilitate auto-discovery of ATM connections by an NMS. An IME can retrieve these objects from the MIB instance of its neighboring IME via ILMI protocol; the IME makes the values available to the NMS of this ATM device through its network management MIB. The NMS, in turn, may retrieve, via an SNMP protocol, the ILMI MIB of the neighboring ATM device for its adjacency information using the atmfMyIpNmAddress obtained as the IP address of the target system. By repeating this process, the NMS can find out the topology of the network it manages and outside connections of the network. The AtmfPortMyIfName can be used to distinguish the parallel links between the same two ATM devices to avoid any topological ambiguity. Chapter 6 discusses an example of topology discovery using adjacency information.

5.2.1.4 ATM Layer Interface

The management of the ATM layer interface involves two groups of objects in the ILMI MIB, the per-ATM layer interface group, which provides general information about the interface; and the per-ATM layer interface statistics group, which contains performance statistics specific to the ATM layer interface.

5.2.1.4.1 Per-ATM Layer Interface Group

This group contains per-ATM layer interface configuration information that can be classified into two categories, namely VPC/VCC configuration and protocol information, as listed in Table 5.3. The ATM layer interface is identified by the interface index (atmfAtmLayerIndex).

Table 5.3
Objects in the ATM Layer Interface Group

Prefix	atmfAtmLayer		
Category	Name	Definition	Comments
Index	Index	The same as that for the physical interface	
VPC/VCC configuration information	MaxVpiBit MaxVciBitss	Maximum number of VPI/VCI bits that may be active	
	MaxVPCs MaxVCCs	Maximum number of VPCs or VCCs, including both SVCs and PVCs, that the local interface can support	
	MaxSvpcVpi MaxSvccVpi	Maximum VPI that the signaling stack on the ATM interface is configured to support for allocation to switched VP/VCCs	
	MinSvccVci	Smallest VCI value that the signaling stack is configured to support for allocation to switched VCCs; the value applies to all SVCC VPI values for which the signaling stack is configured	
	ConfiguredVPCs ConfiguredVCCs	Current number of permanent VPCs/VCCs for which the local interface is configured to process	Number of entries in the atmfVpcTable/ atmfVccTable
Protocol information	DeviceType	The type of the ATM device (i.e., either public or private)	

Prefix	atmfAtmLayer		
Category	Name	Definition	Comments
	UniType	An indication of the latest version of the ATM Forum signaling specification that is supported on this ATM interface	
	UniVersion	The latest version of the ATM Forum UNI Specification that is supported on this ATM Interface.	
	IlmiVersion	The latest version of the ATM Forum ILMI specification that is supported on this ATM interface	
	NniSigVersion	The latest version of the ATM Forum PNNI specification that is supported on this ATM interface	PNNI routing version is not determined through ILMI

VPC/VCC configuration information. Objects in the VPC/VCC configuration category contain the information regarding the capability of the ATM switching fabric in an ATM switch. For a detailed description of ATM switching fabrics, please refer to Chapter 3. It should be noted that the division of ATM switching fabrics into VPC/VCC switches and cross-connects does not necessarily mean that an ATM switch physically has two types of switching fabrics. Typically, only one fabric is used for both switched connections and permanent connections. However, ATM switches usually do reserve different ranges of VPC/VCC numbers for switched and permanent connections, respectively. Moreover, a certain ATM switch may only support a certain range of VPI/VCI numbers due to limitations of its implementation—for instance, the capacity of the switching matrix. To ensure interoperability, ILMI provides a mechanism for ATM switches at both ends of an interface to negotiate a set of mutually supported parameters.

The allocation of VPI/VCI is depicted in Figure 5.5. The switched VPCs for an ATM interface start from a single contiguous range of VPIs beginning with VPI = 1, (VPI = 0 is used for ILMI and signaling VCCs) and ending with the number indicated by the maximum switched VPC VPI object (atmfAtmLayerMaxSvpcVpi). Although permanent VPCs may be allocated anywhere, ILMI suggests that permanent VPCs be given VPI values greater than the maximum SVPC VPI and less than the maximum VPI.

Again, the switched VCCs for an ATM interface are allocated from a contiguous range of VCIs beginning with the number indicated by the minimum switched VCC VCI object (atmfAtmLayerMinSvccVci) and ending with the maximum VCI. The same value applies to all switched VCC VPI values for which the signaling stack is configured. To be compliant with ATM Forum specifications for well-known virtual circuits, this value should be at least 32. Permanent VCCs may be allocated anywhere, but it is suggested that permanent VCCs be given VCI values less than the minimum SVCC VCI.

Because of hardware and software limitations, the maximum number of VPCs/VCCs is less than or equal to two raised to the power of the maximum number of active VPI/VCI bits.

Adjacent IMEs negotiate VPC/VCC configuration parameters at ILMI initialization. Each IME retrieves the values of the *maximum* objects from its

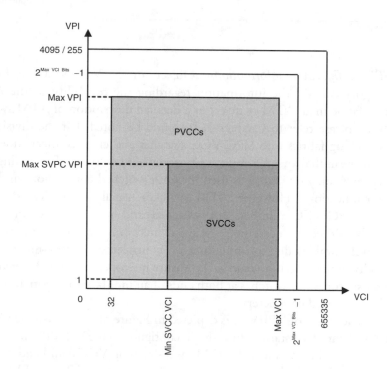

Figure 5.5 Switched VCC VPI and VCI values. (*Copyright 1996 The ATM Forum.*)

peer, including the maximum number of active VPI/VCI bits, maximum number of VPCs/VCCs, and maximum SVPC VPI / SVCC VPI. The active VPI/VCI bits identify the bits in the VPI/VCI fields that are used for cell switching. For more information please refer to Chapter 3. The retrieved value is compared with the local value for each object. The actual value used is set to the lower of the two in order to guarantee interoperability. Similarly, each IME retrieves the value of the minimum switched VCC VCI object. The actual value used is set to the higher one.

Protocol information. With protocol information objects defined in this category, ILMI is capable of automatically identifying the connection type and signaling protocols supported by the peer entity. The information includes the ATM interface type that identifies whether it is public or private, the device type that indicates if it is a user device or a network node; and the latest versions of the ILMI/UNI/NNI protocols supported at the interface.

Adjacent IMEs also negotiate suitable protocols that are supported by both entities. If the peer IME's value of any protocol version object is the same as, or later than the local IME's value, then the version corresponding to the local IME's value should be attempted. Otherwise, if the peer IME's value of the object is earlier and supported locally, then the local IME should attempt the version corresponding to the peer IME's value. This procedure ensures the compatibility of protocols used between adjacent IMEs.

One example of the use of this information is the automatic configuration of protocol for UNI signaling. Unfortunately, there is an incompatibility in the link layer of the signaling protocol between versions 3.0 and 3.1, which prevents interoperability. The UNI Version object in this MIB allows each IME to inspect the highest version of the UNI specification supported by the other. Thus, a system that supports both 3.0 and 3.1 can automatically configure itself to interoperate with its neighbor.

5.2.1.4.2 *Per-ATM Layer Interface Statistics Group*

This group contains per-ATM Layer interface statistics. Because standard network management MIBs such as MIB II interface group already provide such information (as described in Chapter 2), this group has now been deprecated and should not be implemented except as required for backward compatibility. The objects defined in this group are listed in Table 5.4.

Table 5.4
Objects in the Per-ATM Layer Interface Statistics Group

Prefix	atmfAtmStats		
Category	Name	Definition	Comments
Interface index	Index	The same as that for the physical interface	
Statistics	ReceivedCells	The number of received cells on this UNI that were assigned and not dropped	
	DroppedReceived-Cells	The number of received cells dropped due to an unrecoverable header error, invalid cell header, or unknown VPI/VCI values	
	TransmittedCells	The total number of transmitted cells across this UNI interface.	

5.2.1.5 Permanent Virtual Link Management

ATM supports two types of virtual connections, the VPC and VCC. The scope of ILMI connection management is limited to permanent virtual links, since ILMI only manages interfaces between adjacent IMEs. As already described in Chapter 3, a virtual link is identified by a link identifier and characterized by a set of traffic and QoS parameters.

Virtual link management involves four groups of objects defined in the ILMI link management MIB. The per-VP attributes group and the per-VP ABR attributes group are defined for VPL management. The per-VP attributes group contains common information for managing a virtual connection; while the per-VP ABR attributes group contains additional objects specific to support ABR service. Similarly, the per-VC attributes group and the per-VC ABR attributes group are defined for VCL management.

Both VP and channel management follow the same model in which information defined for managing a virtual link can be classified into three categories as follows:

1. Index: Indices to identify an entry of this virtual link in the VP/channel table include the port index, which identifies the ATM port over which this virtual link is set up, and the virtual link identifier, which identifies the virtual link for a given ATM port. For a VPL, this identifier is its VPI, while for a VCL, the identifier includes both the VPI and VCI.

2. Status: Operational status indicating the current state of this virtual connection: unknown, end2endUp, end2endDown, local- Up-End2endUnknown, or localDown.

3. Configuration: Configuration information for this virtual link. Objects defined in this category conform to the ATM Forum's traffic management specification version 4.0 [3], in which a set of traffic, QoS, and ABR operational parameters is specified. These objects include the transmit/receive traffic descriptors, transmit/receive QoS classes, best effort indicator, service category, ABR initial cell rate, and ABR maximum number of data cells per forward RM-cell. For a detailed description of these objects, please refer to Chapter 12. In consideration of the fact that not all ATM switches today support ABR, objects in this category are organized into two groups: the per-VP attributes group and the per-VP ABR attributes group. As a result, those switches that do not support ABR may choose to implement the first group only.

Note that the ILMI link management MIB does not define any objects for per-virtual link performance monitoring. Tables 5.5 and 5.6 list the objects defined for managing a VP in the per-VP attributes group and per-VP ABR attributes group, respectively. Similar objects are also used to manage a VC. However, the management of VCC is slightly different from that of VPC in that additional objects are specified, as listed in Table 5.7. In particular, VCI is added to index an entry in the VCC table; and VCC-specific ABR attributes such as the transmit/receive frame discard indication objects are also included to configure a VCC.

Table 5.5
Objects in the Per-VP Attributes Group

Prefix	atmfVpc		
Category	Name	Definition	Comments
Index	PortIndex	The same as that for the physical interface	
	Vpi	VPI value for this VPC	
Status	OperStatus	Operational status of this VPC	
TX Traffic Descriptor	TransmitTrafficDescriptorType	Type of traffic that is received from this VPC	
	TransmitTrafficDescriptorParam1 - 5	The first to the fifth parameters of the transmit traffic descriptor	
	TransmitQoSClass	QoS class as specified in UNI 3.1	Deprecated
Rx Traffic Descriptor	ReceiveTrafficDescriptorType	Type of traffic that is received over this VPC	
	ReceiveTrafficDescriptorParam1 - 5	The first to the fifth parameters of the receive traffic descriptor	
	ReceiveQoSClass	QoS class as specified in UNI 3.1	Deprecated
Service Indication	QoSCategory	For backward compatibility with version 2.0 of the UNI specification only	Deprecated
	BestEffortIndicator	Indicator specifies whether best effort is requested for this VPC	
	ServiceCategory	Service category of this VPC	

Table 5.6
Objects in the Per-VP ABR Attributes Group

Prefix	atmfVpcAbr		
Category	Name	Definition	Comments
Index	PortIndex	The same as that for the physical interface	
	Vpi	VPI value for this VPC	
ABR operational parameters	TransmitIcr	Initial cell rate	Refer to TM 4.0 for further details
	TransmitNrm	Maximum number of data cells per forward RM-cell	
	TransmitTrm	Maximum time between forward RM-cells	
	TransmitCdf	Cutoff decrease factor	
	TransmitRif	Rate increment factor	
	TransmitRdf	Rate decrease factor	
	TransmitAdtf	ACR decrease time factor	
	TransmitCrm	RM-cells before cutoff	

Table 5.7
Additional Objects Used for VCC Management

Prefix	atmfVpcAbr		
Category	Name	Description	Comments
Additional Index	Vci	VCI value for this VCC	
Frame discard indication	TransmitFrameDiscard	Transmit frame discard indication	
	ReceiveFrameDiscard	Receive frame discard indication	

5.2.1.6 Link Management Traps

Two traps are defined for the ILMI in order to indicate a newly configured, modified, or deleted permanent VPC or permanent VCC. The variables included in the traps uniquely identify the VPI or VPI/VCI values of the reconfigured VPC or the VCC at this ATM interface. Table 5.8 describes these two traps.

5.2.2 Address Registration MIB

Address registration, an important feature of ILMI, greatly enhances the administration of ATM addresses. ILMI supports address registration when used at a UNI, providing a mechanism for automatic exchange of addressing information across a UNI. It uses SNMP over UNI links to access an ILMI MIB associated with the link

The private AESA, as described in Chapter 3, is divided into two parts: a 13-octet network prefix consisting of IDP and HO-DSP and a seven-octet user part including ESI and SEL. In address registration, an ATM end system on the user side of a UNI obtains and registers its address automatically by combining the network prefix that is provided by the network-side IME and the ESI that is provided by the user-side IME. The network-side IME is allowed to supply multiple network prefixes for use at a single UNI; however, it is expected that just one will normally be supplied.

The ILMI address registration MIB consists of three groups: the network prefix group, which contains network prefixes; the address group, which contains registered ATM addresses; and the address registration admin group

Table 5.8
ILMI Traps

Trap Name	Parameters	Description
atmfVpcChange	PortIndex, Vpi and Status	Indicates that a permanent VPC has added or deleted at this ATM interface or that the attributes of an existing VPC have been modified
atmfVccChange	PortIndex, Vpi, Vci and Status	Indicates that a permanent VCC has added or deleted at this ATM interface or that the attributes of an existing VCC have been modified

indicating whether address registration is supported at the UNI. The structure of this MIB is depicted in Figure 5.6.

5.2.2.1 Network Prefix Group

The network prefix group must be implemented by the IME on the user side of the private UNI and may be implemented by the IME on the user side of the public UNI. This group consists of one table, the network prefix table, which is indexed by the ATM interface index and by the value of a network prefix. The objects in this group are listed in Table 5.9. Network-side IME supplies a network prefix by creating an entry in the network prefix table in the user-side IME.

5.2.2.2 Address Group

The address group must be implemented by the IME on the network side of the private UNI and may be implemented by the IME on the network side of the public UNI. This group consists of one MIB table, the address table, which is indexed by the ATM interface index and by the value of an ATM address. The objects in this group are listed in Table 5.10. User-side IME supplies an ATM address by creating an entry in the network address table in the network-side IME.

The organizational scope object is defined to support scope for individual and group addresses. This object is required to inform the network of the desired scope of registration, which in turn can be used in PNNI or to support anycast capability in UNI 4.0[4]. A brief description of the scope is listed in Table 5.11.

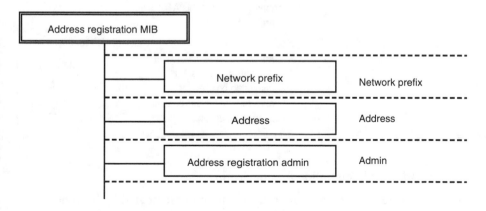

Figure 5.6 Structure of the address registration MIB.

Table 5.9
Objects in the Network Prefix Group

Prefix	atmfNetPrefix		
Category	Name	Description	Comments
Index	Port	The same as that for the physical interface	
	Prefix	Network prefix	
Status	Status	Indicates the validity of a network prefix at this ATM Interface	

Table 5.10
Objects for the Address Group

Prefix	atmfAddress		
Category	Name	Description	Comments
Index	Port	The same as that for the physical interface	
	AtmAddress	The ATM address on the user side of the ATM UNI port	
Status	Status	Indicates the validity of a network prefix at this ATM interface	
Scope	OrgScope	The ATM address organizational scope for the associated address; this object can be used for individual addresses as well as group addresses	Used in PNNI or anycast

5.2.2.3 Address Registration Admin Group

The address registration admin group must be implemented by both network-side and user-side IMEs. This table is used by a local IME to indicate to its peer

Table 5.11

Organizational Scope

Scope	Description
1	local network
2	localPlusOne
3	localPlusTwo
4	siteMinusOne
5	intraSite
6	intraSitePlusOne
7	organizationMinusOne
8	intraOrganization
9	organizationPlusOne
10	communityMinusOne
11	intraCommunity
12	communityPlusOne
13	Regional
14	interRegional
15	Global

IME whether it supports the prefix or address group. This group consists of one MIB table, the address registration admin table, indexed by the ATM interface index. The status object indicates whether address registration is supported at the IME. This object is always set to support at private UNIs, because the prefix and address groups are mandatory at the interfaces. The objects in this group are listed in Table 5.12. Note that the status object is read-only, reflecting the fact that this table is configured by the agent.

5.2.2.4 Address Registration

Address registration can be performed at or after ILMI initiation by invoking SNMP protocol operations. A simplified procedure for address registration at initialization is depicted in Figure 5.7.

Table 5.12
Objects in the Address Registration Admin Group

Prefix	atmfAddressRegistrationAdmin		
Category	Name	Description	Comments
Index	Index	The same as that for the physical interface	
Status	Status	Indicates the support of the prefix and address groups	

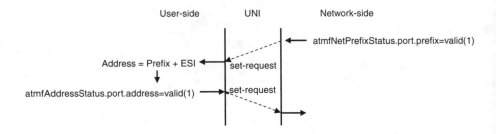

Figure 5.7 Address registration.

At initialization, the registration of network prefixes occurs first. The network-side IME supplies network prefixes for the user-side IME by issuing set-request messages to create entries in its peer's ATM network prefix table in order to register network prefixes.

Subsequently, the user-side IME combines each of the user parts (ESIs) it wishes to use with one or more of the registered network prefixes to form a set of ATM addresses. The user-side IME then registers these addresses by sending set-request messages to create entries in the ATM network prefix table in the network-side IME in order to register network addresses.

After initialization, the network-side IME issues set-request messages to create/delete entries in the network prefix table as and when new network prefixes need to be added or existing network prefixes need to be deleted. Similarly, the user-side IME issues set-request messages to create/delete entries in the address table as and when new ATM addresses need to be registered or existing ATM addresses deregistered. If and when the IME loses ILMI connectivity, all addresses are deregistered.

The following example illustrates the sequence of the address registration, as shown in Figure 5.7. The network-side IME issues an SNMP set-request to set an entry in the network prefix table at the user-side. Assume that the network prefix of the NSAP for the switch is 71.0.5.128.255.225.0.0.0.242.21.20.167. Note that the first '0' is the port index. The remaining 14 octets consist of the prefix encoded as OCTETSTRING, the first octet of which is the length of the string (i.e., 13).

```
set-request (
atmfNetPrefixStatus.0.13.71.0.5.128.255.225.0.0.0.242.21.20.167 = valid(1) );
```

The user-side IME combines the network prefix received and its MAC address (00.32.72.16.20.43) to form its NSAP address. Then, it issues a set-request to create an entry of the address in the address table at the network side with parameters below. Note that the last '0' is the ESL field. Again, the first '0' is the port index, while the octet that follows it is the length of the NSAP address, i.e., '20.'

```
set-request (
atmfAddressStatus.0.20.71.0.5.128.255.225.0.0.0.242.21.20.167.0.32.72.16.20.43.0
= valid(1) );
```

5.2.3 Service Registry MIB

The service registry MIB offers a general-purpose service registry mechanism for locating ATM network services such as the LECS. Descriptions of the objects in this group can be found in Table 5.13.

Here is an example of retrieving ATM network services using the MIB (this may start with an SNMP get-next-request):

```
get-next-request ( atmfSrvcRegATMAddress ) ;
```

The response could resemble the following:

```
response (
  atmfSrvcRegATMAddress.0.atmfSrvcRegLecs.1
    = 20.71.0.5.128.255.225.0.0.0.242.21.20.167.0.32.72.16.20.43.0 );
```

The response contains information of one entry in the table including service type (LANE) and the address from which the service is provided.

Chapter 9 describes how to use LECS to provide LANE service.

Table 5.13
Objects in the service registration MIB

Prefix	atmfSrvcReg		
Category	Name	Description	Comments
Index	Port	The same as that for the physical interface	
	ServiceID	Service identifier to identify the type of service provided	
	AddressIndex	An arbitrary integer to differentiate multiple entries containing different ATM addresses for the same service on the same port	
Address	ATMAddress	Full ATM address of the service to which the user-side IME may establish a connection with the service	
Parameter	Parm1	Parameter that may be used for the service supported	

References

[1]   The ATM Forum Technical Committee, "User-Network Interface (UNI) Specification, Version 3.1," September 1994.

[2]   ATM Forum, "Integrated Local Management Interface (ILMI) Specification Version 4.0," af-ilmi-0065.000, September 1996.

[3]   The ATM Forum Technical Committee, "ATM Forum Traffic Management Specification, Version 4.0," April 1996.

[4]   The ATM Forum Technical Committee, "ATM User-Network Interface (UNI) Signalling Specification, Version 4.0," July 1996.

6

AToM MIB

6.1 Overview

RFC 1695 [1], "Definitions of Managed Objects for ATM Management Version 8.0," specifies an MIB for the management of ATM networks. This MIB is also known as the AToM MIB, named after the IETF working group that developed it. The same working group is responsible for defining the MIB to manage SONET. AToM is actually a combination of two parts: ATM and a lowercase "o" which signifies SONET. As a whole, it is a reminder to keep the MIBs small and efficient. The resulting ATM MIB, though not small, is well within control, given the complexity of ATM management. The current status of this MIB on the standard track is *proposed standard*. The SONET MIB will be introduced in Chapter 7.

The AToM MIB defines objects to manage ATM interfaces, virtual links, cross-connects, and AAL5 entities and connections supported by ATM hosts, ATM switches, and ATM networks. It is both compliant to the SNMPv2 SMI, and semantically identical to the peer SNMPv1 definitions and can thus be accessed by both the SNMPv1 and the SNMPv2 management applications.

The managed objects defined in the AToM MIB can also be used for service management. For example, this MIB is specified to represent a combination of switches (i.e., an ATM network) for CNM as introduced in Chapter 4.

The primary purpose of this MIB is to manage ATM permanent virtual connections including PVPCs and PVCCs. Although SVPCs/SVCCs can be

155

represented in the MIB, full management of switched connections requires additional capabilities that are beyond the scope of the AToM MIB.

6.2 Management Models

This section briefly introduces management models in the ATOM MIB.

6.2.1 Permanent Virtual Connection Management

End-to-end permanent virtual connection management is the key to the management of an ATM network. A single ATM end system or switch does not support the whole end-to-end span of a virtual connection (VPC or VCC). Rather, a virtual connection traverses multiple ATM end-systems and/or switches, each supporting one piece of the connection (i.e., a virtual link). That is, each ATM end system at one end supports its end of virtual connection plus the virtual links on its external interfaces, while each *intermediate system* (IS) (i.e., ATM switch) through which the VC passes, supports the multiple virtual links on that switch's external interfaces and the cross-connection of those virtual links through that switch. Thus, the end-to-end management of a virtual connection is achieved by appropriate management of its individual pieces in combination, which is depicted in Figure 6.1.

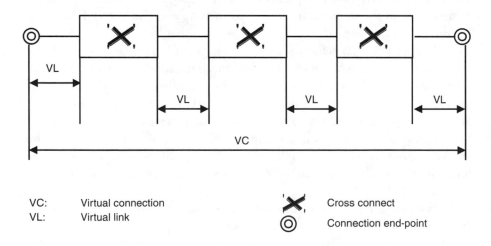

VC:	Virtual connection
VL:	Virtual link

Cross connect

Connection end-point

Figure 6.1 End-to-end virtual connection management.

A virtual connection is associated with a set of traffic descriptors that specifies the traffic characteristics including traffic parameters and QoS class (or the QoS category as used in the recent traffic management specification). Virtual links inherit traffic characteristics from the virtual connection of which they are a part. The traffic parameters in two directions of a connection can be symmetric or asymmetric (i.e., the two directions can have the same or different traffic characteristics). The traffic descriptors used in this MIB are consistent with those used in the UNI 3.0 ILMI MIB defined by the ATM Forum. Please refer to Chapter 12 for a detailed description of ATM traffic models and their representation in relation to SNMP network management.

6.2.2 AAL5 Management

The role of ATM AAL is to map the information transfer protocols onto ATM. The AToM MIB supports the AAL5 management by utilizing different models in ATM hosts and switches.

6.2.2.1 Managing AAL5 in an ATM Switch

AAL5 is managed in a switch for only those virtual connections that carry AAL5 and are terminated at the AAL5 entities inside the switch typically to support the signaling channels. The VCCs within the ATM UNIs carrying AAL5 are switched by the ATM switching fabric (termed an ATM entity) to the VCs on a proprietary internal interface associated with the AAL5 process (termed an AAL5 entity). Therefore, performance management of the AAL5 resource in the switch is modeled using the ifTable through an internal (pseudo-ATM) virtual interface, and the AAL performance management per-virtual connection is supported using an additional AAL5 connection table in the ATM MIB. The association between the AAL5 virtual link at the proprietary virtual internal interface and the ATM virtual link at the ATM interface is derived from the VC cross-connect table and the VC table in the ATM MIB. The management of AAL5 in a switch is depicted in Figure 6.2(a).

6.2.2.2 Managing AAL5 in a Host

Managing AAL5 in a host is straightforward as the VCCs are terminated at the host in which the AAL5 sublayer is stacked directly over the ATM sublayer. The management of the AAL5 sublayer interface in an ATM host is shown in Figure 6.2(b).

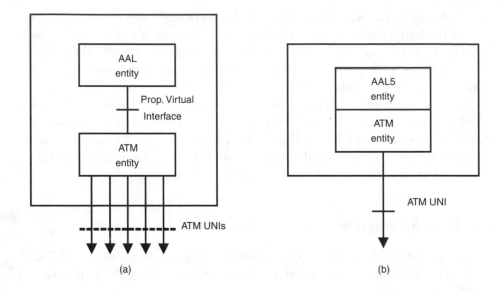

Figure 6.2 AAL5 management (AToM): (a) in a switch and (b) in an ATM host.

6.3 The Structure of the MIB

The structure of the AToM MIB is depicted in Figure 6.3. In addition to the
MIB groups (shown as solid line blocks in the figure) defined in this MIB,
other MIB modules are necessary to manage ATM interfaces, links, and cross-
connects. Examples include MIB II for general system and interface management (RFC 1213 and RFC 1573), the DS3 or SONET MIBs for management
of physical interfaces MIB (shown as dashed line blocks), and, as appropriate,
MIB modules for applications that make use of ATM, such as SMDS, PNNI,
and LANE MIBs. These MIB modules are outside the scope of AToM MIB,
and are, therefore, not shown in Figure 6.3. Each group of related managed
objects is represented in the AToM MIB as a conceptual table that is described
in the SNMPv2 SMI.

6.3.1 Physical Interface

The management of the physical layer of an ATM interface requires appropriate physical layer MIBs, such as the DS3/E3 MIB and the SONET MIB which
will be introduced in Chapter 7. AToM MIB defines two managed objects

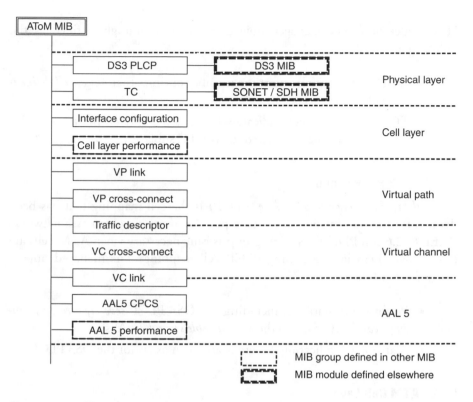

Figure 6.3 AToM MIB structure.

groups, namely the ATM interface DS3 PLCP group and the ATM interface TC sublayer group to manage the physical layer functionality that maps the ATM layer to the underlying SONET or DS3 links over which ATM cells are transmitted. Again, the PLCP table and the TC table conceptually extend the ifTable. Thus, a row in these tables is indexed by the ifIndex, which in turn uniquely identifies the ATM interface which sits above the PLCP and/or TC sublayer. The management of ATM over DS1/E1 is very similar to that of the ATM over DS3/E3, however, this has not been specified in the AToM MIB.

6.3.1.1 TC Sublayer Group

TC sublayer is responsible for embedding the ATM cells in the transmission frames of the transport medium that is in use. This layer is necessary for both the DS3 PLCP and the SONET links where cells are transmitted using a transmission frame. Two of its most important functions are cell delineation and

HEC generation. The managed objects are correspondingly defined as the following:

- OCDEvents, which report the number of times that *out of cell delineation* (OCD) events occur;
- TCAlarmState, which indicate whether a *loss of cell delineation* (LCD) failure state has been declared for the TC sublayer.

6.3.1.2 DS3 PLCP Group

Physical layer convergence procedure (PLCP) is the cell mapping that has been defined for the transmission of the data packets of metropolitan area networks (IEEE 802.6) on PDH lines. This group is only necessary when ATM cells are carried over DS3 interface using PLCP cell mapping. The managed objects defined are as follows:

- Performance statistics, including the DS3 PLCP *severely errored farming seconds* (SEFSs) and the *unavailable seconds* (UASs);
- Alarm state, indicating if there is an alarm exists for the DS3 PLCP.

6.3.2 ATM Cell Layer

An ATM cell layer interface is represented as one entry in the MIB-II ifTable. This entry regards the ATM cell layer as a whole, rather than as individual virtual connections that are managed via the ATM-specific managed objects specified in this MIB, thus avoiding unnecessary ifEntry proliferation and implied performance statistics per VP/VC.

6.3.2.1 ATM Interface Configuration Group

To configure the ATM interface, two groups of objects are used. They are the MIB II interface group and the ATM interface configuration group. The former contains the general interface information, whereas the latter is an ATM-specific conceptual extension of the MIB II ifTable. Each ATM cell layer interface is considered as a whole in AToM MIB; hence, it is represented as only one entry in the MIB II ifTable. The ATM-specific objects defined for an ATM cell layer interface are defined as follows:

- Neighboring information including interface name and IP address of the neighbor to which this interface connects. These addresses can either be manually configured or automatically discovered through ILMI protocol, as discussed in Chapter 5.

- VPC/VCC configuration information including the maximum number of VPCs and VCCs, maximum number of active VPI/VCI bits supported, and the number of VPCs and VCCs that are currently in use at the interface. The object definitions are identical to those defined in the ILMI MIB, as discussed in Chapter 5.
- The VPI and VCI of the ATM VCC connection to transfer the ILMI management information.

Figure 6.4 depicts the roles of the ifTable and the ATM interface configuration table in this group.

6.3.2.2 Cell Layer Performance Group

As mentioned before, the ATM cell layer is considered as a whole; therefore, each ATM interface is represented as only one entry in the ifTable of MIB II. The objects in an entry of this table are used to collect status and performance information of the cell layer. Detailed mapping can be found in the AToM MIB [1].

6.3.3 Virtual Path

Managing VPCs involves three groups in the AToM MIB. The VPL group describes VPLs within an ATM end system or a switch. The traffic descriptor parameter group contains information about VPL/VPC traffic characteristics. Last, the VP cross-connect group includes connection information about the VPCs supported within a switching system.

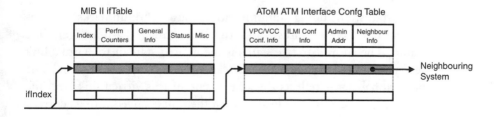

Figure 6.4 MIB tables for ATM interface management.

6.3.3.1 Traffic Descriptor Parameter Group

The traffic descriptor group contains a set of self-consistent ATM traffic parameters, including the QoS class. Each entry in the ATM traffic descriptor parameter table describes a specific type of traffic and consists of two components:

- A vector of traffic descriptors, including an ATM traffic descriptor type column, describing the type of traffic; five columns that parameterize the traffic (atmTrafficDescrParam1 through atmTrafficDescrParam5, the number of columns actually used depends on the traffic type); and a column indicating the QoS class;
- An index (atmTrafficDescrParamIndex) to identify the entry of the table and row status that is used to create, modify and delete a row in the table.

Unlike other tables in this MIB, entries in this table do not belong to any specific interface. Information contained in this group is system-wide, meaning that it can be shared by any interface within the system. As a result, an entry in this table is indexed by the atmTrafficDescrParamIndex. A detailed description of the parameters used in this group can be found in Chapter 12. As of this writing, the AToM MIB group is revising the AToM MIB. It is most likely that this group will be updated according to the traffic management 4.0.

6.3.3.2 Virtual Path Link Group

This group defines necessary attributes to describe a VPL. Each VPL has an entry in the VPL table containing the following information:

- VPL identifier (atmVplVpi), which uniquely identifies the link, that is, the VPI value of the VPL.
- Link status, including administrative status (AdminStatus), operational status (OperStatus), and the last change (LastChange) time stamp. The administrative status is the desired state of the virtual link set by a network manager, which can be either "up" or "down." The operational status reflects the actual state of this virtual link, and the last change timestamp indicates when the link changed into its current state.
- Traffic characteristics, including the traffic descriptor indexes pointing to appropriate entry or entries in traffic descriptor table, which specify

the characteristics of the traffic transmitting over the VP. The VPL is bi-directional; hence there are two traffic descriptors, one for each direction of transmission. The traffic can be symmetric or asymmetric. In the former case, one entry in the traffic descriptor table is sufficient, whereas in the latter, there should be two entries.

- Cross-connect identifier (atmVplCrossConnectIdentifier) is defined to identify a VPC that is point-to-point, point-to-multipoint, or multipoint-to-multipoint. The use of this identifier will be examined in Section 6.3.3.3.

An entry in this table is uniquely identified by the interface index (ifIndex) and VPL index (atmVplVpi).

6.3.3.3 Virtual Path Cross-Connect Group

The VP cross-connect group contains configuration and state information of all point-to-point, point-to-multipoint, and multipoint-to-multipoint VP cross-connections within a switching system. A point-to-point bi-directional VPC is modeled as one entry; a point-to-multipoint as N entries, and a multipoint-to-multipoint VPC cross-connect as $N(N - 1)/2$ entries in the cross-connect table. In the latter two cases, the entire N or $N(N - 1)/2$ entries are associated with the same value of cross-connect index. This group is only necessary in the ATM switches. Each entry in the table contains the following information:

- An index identifying this cross-connection; a read-only object called atmVpcCrossConnectIndexNext is used to generate unique identifiers as discussed in Chapter 2. Each time the object is read, a unique value is returned to the management application as the index for the cross-connection to be created. After each retrieval, the agent modifies the value to the next unassigned index. This guarantees that the index values are unique, even in the case where multiple managers are creating cross-connections simultaneously.

- A reference to both of the virtual links connected by this cross connect; an entry in the cross-connect table is identified by the cross-connect index, the interface indexes, and the VPIs of the two cross-connected VPLs (i.e., the low interface index, high interface index, low VPI, and high VPI). The terms low and high are chosen to represent numerical ordering of the two interfaces within a VPC cross-connect. That is, the

ATM interface with the lower value of ifIndex is termed *low,* while the higher value one is termed *high* interface.

- Status, reflecting the state of the cross-connection including administrative status, low to high operation status, high to low operation status, and the time stamp for the last change in both the low to high and the high to low directions.

6.3.4 Virtual Channel

Managing the VCCs is almost the same as managing VPs, which also involves three groups, the VCL group, the VC cross-connect group, and the traffic descriptor parameter group. In order to avoid repetition, only those objects that are different from those in the VP groups are addressed in Sections 6.3.4.1 and 6.3.4.2.

6.3.4.1 Virtual Channel Cross-Connect Group

Two more objects, the high VCI and low VCI, are added to this group in addition to the objects similar to the VP cross-connect group. The reason is that a VCL is identified by VPI/VCI; in contrast, only VPI is necessary for identifying VPL. As a result, the indexes for an entry in the VC cross-connect table also include the high and low VCIs.

6.3.4.2 Virtual Channel Link Group

For the same reason explained in Section 6.3.4.1, the VCL group includes the VCI value as a reference to access the entry in the VC table. Because AAL layer resides over VC layer, this group defines some additional objects to configure the AAL entity that is attached to a VCC end-point. It defines the AAL-specific objects, including the type of AAL used and for AAL5, the maximum transmit and receive SDU sizes and the type of data encapsulation. The object for the AAL type is only valid when the VCL ends a VCC at the ATM system. The AAL5-related objects are only valid if the type of the AAL used is AAL5.

To summarize VP/VC management, Figure 6.5 depicts how different tables are used to represent a segment of point-to-point virtual connection within an ATM switch. The segment consists of two VPLs/VCLs that are cross-connected by a switch. Each VPL/VCL is represented as an entry in the VPL/VCL table identified by the appropriate ifIndex and its VPI/VCI, which are collectively referred to as link index in the figure. The cross-connect index points to an entry in the VPL/VCL cross-connect table, which in turn contains the low link and high link indexes to identify the entries of high/low

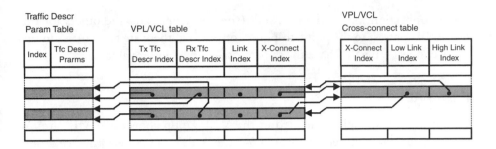

Figure 6.5 Point-to-point connection.

VPL/VCL, respectively. The traffic parameters of the VPC/VCC are maintained in the traffic descriptor parameter table, the entries of which are identified by the transmit/receive traffic descriptor index. Only one entry is used for symmetric traffic—i.e., traffic parameters in both directions are the same. For asymmetric traffic, however, two entries should be used.

Figure 6.6 illustrates an example of a point-to-multipoint connection, which consists of a root VPL/VCL and two leaf VPLs/VCLs. For a point-to-multipoint connection, the traffic is one-directional (i.e., from the root to the leaves), whereas the bandwidth for the reverse direction is zero. Thus, the traffic parameters are contained in only one entry in the traffic descriptor parameter table, which is identified by the receive traffic descriptor index of the root or the transmit indexes of the leaves.

Again, for each virtual link, there is a corresponding entry in the VPL/VCL table. Hence, there are three entries used for this one point-to-two-

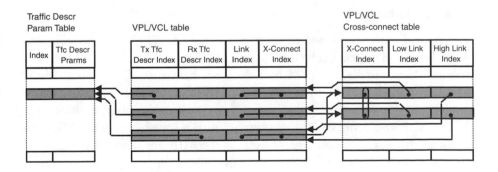

Figure 6.6 Point-to-multipoint connection.

point connection, or for the root link (the lower one) and the others for the leaves. In contrast to the point-to-point connection, there are two entries in the cross-connect table that have the same cross-connect index value.

6.3.5 AAL5

In addition to the objects defined in the VCL group above, two MIB groups are used for the management of AAL5—the AAL5 CPCS performance group in this MIB and the interface group in MIB II. The CPCS group contains the AAL5-specific performance information, while the MIB II interface group provides general information about the AAL5 interface. An AAL5 entity is represented as an independent entry in the MIB II ifTable.

6.5.3.1 AAL5 CPCS Performance Statistics Group

This group contains the following objects to monitor the performance of the AAL5 CPCS (CPCS):

- PDUs discarded for CRC errors;
- PDUs discarded due to reassemble time out;
- PDUs discarded because of long SDUs.

The AToM MIB applies to ATM hosts, ATM switches, and ATM networks. However, a specific ATM device only needs to implement some portions of the MIB depending on the role—end system or switch, for example—it will play.

6.3.6 Conformance

The AToM MIB is designed to support a variety of ATM devices, including ATM hosts and switches. The functionality of these devices differs from one to another. For example, an ATM host does not have the cell switching-fabric. Thus, it is inappropriate to implement the VP/VC cross-connect group. In order to accommodate such requirements while retaining interoperability, the AToM MIB also defines conformance/compliance statements in the MIB as a guide to the implementor. The objects in the MIB are grouped into several groups, from which an ATM device may choose an appropriate set to implement. In addition, the minimum requirements for implementation of each object and object group are specified. (Table 6.1 lists all the conformance

Table 6.1
Conformance Groups for the AToM MIB

Group Name	Description
atmInterfaceConfGroup	Configuration information about an ATM interface
atmTrafficDescrGroup	Information about ATM traffic descriptor type and the associated parameters
atmInterfaceDs3PlcpGroup	Information about DS3 PLCP layer at an ATM interface
atmInterfaceTCGroup	Information about TC sublayer at an ATM interface
atmVpcTerminationGroup	Information about a VPL at an ATM interface that terminates a VPC (i.e., one that is not cross-connected to other VPLs)
atmVccTerminationGroup	Information about a VCL at an ATM interface that terminates a VCC (i.e., one that is not cross-connected to other VCLs)
atmVpCrossConnectGroup	Information about a VP cross-connect and the associated VPLs that are cross-connected
atmVcCrossConnectGroup	Information about a VC cross-connect and the associated VCLs that are cross-connected
aal5VccGroup	AAL5 configuration and performance statistics of a VCC

groups specified in the MIB.) As a result, the host only needs to implement the VPC/VCC termination groups instead of all objects needed to support VPC/VCC management.

6.4 End-to-end Connection Management Using AToM MIB

The AToM MIB focuses on permanent virtual connection management and specifies elaborate procedures to configure, reconfigure, and release VPCs/VCCs; these procedures allow detailed, step-by-step error checking. The MIB treats VCCs and VPCs in much the same way as introduced in Section 6.3. Thus, for the sake of simplicity, procedures are described in the general terms of virtual connections and virtual links. The manipulation of virtual connections can be broken down into several phases, each of which affects different portions of the end-to-end virtual connection.

The end-to-end virtual connection management using AToM MIB will be illustrated with an example of configuring a VCC between host A and host B through one *intermediate system* (IS) (i.e., an ATM switch) as shown in Figure 6.7. This example is based on the one Tesink and Brunner [2] published in *The Simple Times* with some elaboration, modification, and minor corrections, so as to fit into this book. It assumes that the underlying VPLs have already been established.

6.4.1 Virtual Connection Establishment

Virtual connection establishment consists of the following phases:

1. Reserve appropriate virtual links;
2. Characterize traffic on the virtual links.
3. Cross-connect the virtual links in ISs and associate the virtual links with a user application in the hosts.

6.4.1.1 Reserve Appropriate Virtual Links

The management application creates a virtual link entry in the virtual link table (atmVclTable in our example) by setting the row status to createAndWait(5). Note that the VC indices are chosen by the manager, and not by the agent. The index clause of the atmVclTable is {ifIndex, atmVclVpi, atmVclVci}. The network manager starts to reserve virtual links along the route by sending SNMP set-requests to each ATM device involved.

If no error is returned, the agent creates a row and reserves the VPI/VCI values on that port. It also increments the counters of VPCs/VCCs (i.e., atmInterfaceVpcs/Vccs). The interactions between the manager and the agents are shown in Table 6.2.

6.4.1.2 Characterize Traffic on the Virtual Links.

The virtual link tables characterize the traffic for transmit and receive direction by pointing to the appropriate entries in the atmTrafficDescrParamTable. Multiple virtual links in the atmVclTable/atmVplTable can point to the same vector in the atmTrafficDescrParamTable. This technique allows the agent to predefine self-consistent traffic vectors in the atmTrafficDescrParamTable. In addition, the agent may support read-create access, allowing the manager to specify additional vectors.

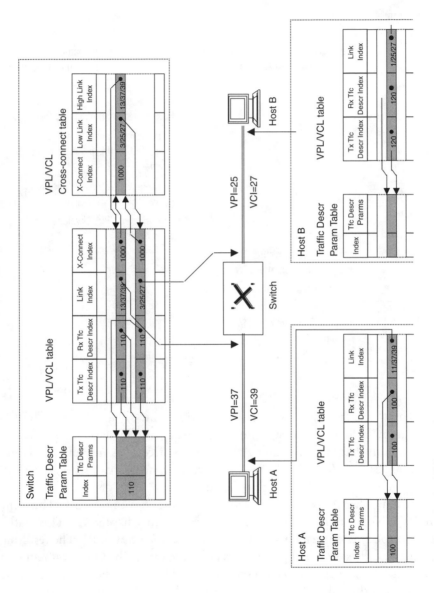

Figure 6.7 Configuration for VCC example.

Table 6.2
Reserving Virtual Links

Destination	Command(s)	Possible causes of errors
Host A	set-request (atmVclRowStatus. 11.37.39) = createAndWait(5));	The agent supports read-only operations on this table only (i.e., (re)configuration is not allowed);
		The selected ifIndex value does not exist, or is not an ATM interface;
		The maximum number of VCs supported for this interface has been reached;
		The selected VPI/VCI values are unavailable for use;
		The selected VPI/VCI values are in use or reserved.
IS	set-request (atmVclRowStatus.13.37.39 = createAndWait(5));	
	set-request (atmVclRowStatus.3.25.27 = createAndWait(5));	
Host B	set-request (atmVclRowStatus.1.25.27 = createAndWait(5));	

Thus, the manager selects an existing row(s) in the atmTraffic-DescrParamTable, or, if no suitable row(s) exists, the manager must create a new row(s) in that table. Assuming that the next free rows in Host A, IS, and Host B are 100, 110, and 120, respectively, the manager can send an SNMP set-request to each device as listed in Table 6.3.

Assume that the VCC is to carry 64-Kbps CES using AAL 1. Thus, the atmNoClpNoScr type should be used according to Chapter 12. This traffic type only specifies one parameter, that is, the *peak cell rate* (PCR). The resulting PCR is 171 cells/s (64,000 bps/47 bytes per cell 8 bits). The QoS class required for the CES is Class A, which corresponds to 1. The commands to set up traffic parameters in the traffic descriptor parameter table are listed in Table 6.4.

Note that only one entry in the traffic descriptor parameter table is required in each system, as the traffic in both directions is symmetric. The management application now characterizes the traffic parameters of all the

Table 6.3
Row Creation in the Traffic Descriptor Parameters Table

Destination	Command(s)	Possible Causes of Errors
Host A	set-request (atmTrafficDescrRowStatus.100 = createAndWait(5));	The agent does not support read-create on this table because only a fixed set of traffic characteristics are supported, for instance; The specified row is already active.
IS	set-request (atmTrafficDescrRowStatus.110 = createAndWait(5));	
Host B	set-request (atmTrafficDescrRowStatus.120 = createAndWait(5));	

virtual links associated with the virtual connection by pointing the receive and transmit traffic index (atmVpl/VclReceiveTrafficDescrIndex and atmVpl/VclTransmitTrafficDescrIndex) in the virtual link table (atmVpl/VclTable) to the atmTrafficDescrParamTable rows containing desired ATM traffic parameter values. The requests issued are listed in Table 6.5.

The manager now activates the virtual links by setting the row status (atmVpl/VclRowStatus) to active, hence the commands in Table 6.6.

If the set-requests are successful, the agent has reserved the resources to satisfy the requested traffic parameter values and the QoS Class for that virtual link.

6.4.1.3 Cross-Connect Virtual Links in the Intermediate Systems

In the IS, the atmVcCrossConnectTable must be used to cross-connect the virtual links. The atmVpl/VclTables have a cross-connect identifier column for this purpose (atmVpl/VclCrossConnectIdentifier). Different rows in the atmVpl/VclTable that have the same cross-connect identifier value are cross-connected. This is achieved through cross-connect tables (atmVp/VcCrossConnectTable).

Before creating a row in the cross-connect table, a unique cross-connect index must be obtained by using atmVp/VcCrossConnectIndexNext. A

Table 6.4
Traffic Parameter Settings

Destination	Command(s)	Possible Causes of Errors
Host A	set-request (atmTrafficDescrType.100 = atmNoClpNoScr);	The parameters are mutually inconsistent;
	set-request (atmTrafficDescrParam1.100 = 171);	The agent does not support the requested values.
	set-request (atmTrafficQoSClass.100 = 1);	
	set-request (atmTrafficDescrRowStatus.100 = active(1));	
IS	set-request (atmTrafficDescrType.110 = atmNoClpNoScr);	
	set-request (atmTrafficDescrParam1.110 = 171);	
	set-request (atmTrafficQoSClass.110 = 1);	
	set-request (atmTrafficDescrRowStatus.110 = active(1));	
Host B	set-request (atmTrafficDescrType.120 = atmNoClpNoScr);	
	set-request (atmTrafficDescrParam1.120 = 171);	
	set-request (atmTrafficQoSClass.120 = 1);	
	set-request (atmTrafficDescrRowStatus.120 = active(1));	

get-next will obtain a value, such as 3333. After retrieval, the agent will increment the value to the next unassigned one. This operation fails if the value 0 is returned, which means that all available values are in use (e.g., the switch can not support more virtual connections).

Cross-connecting virtual connections consists of the following steps:

1. Creating a row in the cross-connect table;
2. Filling in the cross-connect index value in the corresponding virtual link table rows;
3. Activating the row in cross-connect table;
4. Turning on the traffic.

Table 6.5
Reserving Bandwidth for Virtual Links

Destination	Command(s)	Possible Causes of Errors

Host A	set-request (atmVclReceiveTrafficDescrIndex.11.37.39 = 100);
	
	set-request (atmVclTransmitTrafficDescrIndex.11.37.39 = 100);
		Insufficient resources, such as bandwidth and memory

IS	set-request (atmVclReceiveTrafficDescrIndex.13.37.39 = 110);
	
	set-request (atmVclTransmitTrafficDescrIndex.13.37.39 = 110);
	
	set-request (atmVclReceiveTrafficDescrIndex.3.25.27 = 110);
	
	set-request (atmVclTransmitTrafficDescrIndex.3.25.27 = 110);	

Host B	set-request (atmVclReceiveTrafficDescrIndex.1.25.27 = 120);
	
	set-request (atmVclTransmitTrafficDescrIndex.1.25.27 = 120);	

Table 6.6
Activating Virtual Links

Destination	Command(s)	Possible Causes of Errors
Host A	set-request (atmVclRowStatus.11.37.39 = active(1));	Not enough resources

IS	set-request (atmVclRowStatus.13.37.39 = active(1));
	
	set-request (atmVclRowStatus.3.25.27 = active(1));	

Host B	set-request (atmVclRowStatus.1.25.27 = active(1));	

The atmVcCrossConnectTable is indexed jointly by:

```
{
atmVcCrossConnectIndex,
atmVcCrossConnectLowIfIndex,
atmVcCrossConnectLowVpi,
atmVcCrossConnectLowVci,
atmVcCrossConnectHighIfIndex,
atmVcCrossConnectHighVpi,
atmVcCrossConnectHighVci
}
```

Note that cross-connecting VPLs works in the same way, except that there are no VCIs in the INDEX clause. The virtual link indexes for the interface with the lower index value (i.e., ifIndex = 3) must be specified first in the table's index. The commands that need to be sent to the IS are listed in Table 6.7.

At this point, the traffic flow must actually be turned on. In the hosts, the link to the application must be made, which is demonstrated in Table 6.8.

Finally, the manager turns on the traffic at the hosts by issuing the commands listed in Table 6.9.

The above step-by-step process can be shortened by using the createAndGo(4) value for the row-status objects. However, the advantage of detailed step-wise error checking would be lost. Therefore, the step-wise process is recommended.

6.4.2   Graceful Virtual Connection Release

Virtual connection release consists of the following phases:

- Release cross-connect in ISs;
- Release all virtual links.

6.4.2.1   Release the Cross-Connects in the IS

To release a virtual connection, all cross-connects and associated virtual links must be released by setting the row status of the associated table entries to destroy as shown in Table 6.10.

Table 6.7
Cross-Connecting Virtual Links in the IS

Destination	Command(s)	Possible Causes of Errors

IS	1. Create a row in the cross-connect table
	
	set-request (atmVcCrossRowSta-tus.3333.3.25.27.13.37.39 = createAndWait(5));
	
	2. Fill in the cross-connect index value in the corresponding virtual link table rows
	
	set-request (atmVclCrossConnectIdentifier.13.37.39 = 3333);
	
	set-request (atmVclCrossConnectIdentifier.3.25.27 = 3333);
	
	3. Activate the row in cross-connect table
	
	set-request (atmVcCrossConnectRowSta-tus.3333.3.25.27.13.37.39 = active(1));
	
	4. Turn the traffic
	
	set-request (atmVcCrossConnectAdminSta-tus.3333.3.25.27.13.37.39 = up(1));
		The specified low-index value is higher than that of the high-index;
		
		The requested topology is not supported by the agent (e.g., a multipoint VC may be requested that is not supported by this IS);
		
		The traffic and QoS parameter values of the specified VCLs are mutually incompatible;
		
		The IS may be unable to connect the two VCLs due to no network resources being available.

Table 6.8
Associating Virtual Connection Endpoints to the Application

Destination	Command(s)	Possible Causes of Errors
Host A	set-request (atmVclAalType.11.37.39 = aal1(1));	The requested application and virtual link traffic pattern do not match
Host B	set-request (atmVclAalType.1.25.27 = aal1(1));	

Table 6.9
Turning on the Traffic at the Hosts

Destination	Command(s)	Possible Causes of Errors
Host A	set-request (atmVclAdminStatus.11.37.39 = up(1));	
Host B	set-request (atmVclAdminStatus.1.25.27 = up(1));	

Table 6.10
Release the Cross-Connects in the IS

Destination	Command(s)	Possible Causes of Errors
IS	set-request (atmVcCrossConnectRowStatus.3333.3.25.27.13.37.39 = destroy(6));	

This will free the value 3333 for future use by the atmVcCrossConnect-IndexNext, and the atmVclCrossConnectIdentifier values will be removed from the associated virtual links. As a result, cross-connect resources are released.

6.4.2.2 Release the Virtual Links

To reclaim the virtual links associated with the virtual connection, each associated table entry must be destroyed. Hence, the commands listed in Table 6.11 are necessary.

Upon these actions, the agents will release the associated virtual link resources and decrement atmInterfaceVpcs/Vccs.

It is recommended to release a cross-connect before releasing the individual virtual links. The reason is that releasing a virtual link first may, in some implementations, be interpreted as a request for a configuration change (for example, a multipoint connection where one leaf is being deleted). Proper agent implementation should release cross-connects automatically if the following are true.

Table 6.11
Release the Virtual Links

Destination	Command(s)	Possible Causes of Errors
Host A	set-request (atmVclRowStatus.11.37.39 = destroy(6));	
IS	set-request (atmVclRowStatus.13.37.39 = destroy(6));	
	set-request (atmVclRowStatus.3.25.27 = destroy(6));	
Host B	set-request (atmVclRowStatus.1.25.27 = destroy(6));	

- A virtual link is released, and cross-connect reconfiguration is not supported by the agent;
- A virtual link is released and the remaining topology is meaningless to the agent (e.g., one of two cross-connected virtual links is released).

6.4.2.3 Release the Traffic Descriptors

To release the traffic parameter values associated with transmit and receive directions of the virtual links, the rows of the traffic descriptor table (atmTrafficDescrParamTable) pointed to by the virtual links must be deleted. Deletion proceeds in the normal way with the commands in Table 6.12.

6.4.3 Virtual Connection Reconfiguration

Several VC reconfiguration applications are detailed in Section 6.4.3.1 and 6.4.3.2. Some require additional capabilities that an agent may support.

6.4.3.1 Traffic and/or QoS Parameter Value Changes

These do not require additional agent capabilities. The manager takes down the current virtual connections, defines new virtual links with the desired parameters, and brings up the new VC following the rules outlined above. This is most simply done as an entirely new set of virtual links. If there is a desire to retain the VPI/VCI values, the manager may follow these steps:

Table 6.12
Release the Traffic Descriptors

Destination	Command(s)		Possible Causes of Errors
Host A	set-request (atmTrafficDescrRowStatus.100 = destroy(6));		The agent does not support read-write access to this table;
IS	set-request (atmTrafficDescrRowStatus.110 = destroy(6));		The traffic parameters of this row are still used by another VC.
Host B	set-request (atmTrafficDescrRowStatus.120 = destroy(6));		

1. Turn virtual link traffic off at hosts by setting the atmVclAdminStatus to down;

2. Release the cross-connect at ISs by setting the atmVcCrossConnectRowStatus to destroy;

3. Turn virtual link traffic off at ISs by setting the atmVclAdminStatus to down;

4. Take the virtual links out of service at the hosts and ISs by setting the atmVclRowStatus to notInService.

Then, as before, configure the virtual links, cross-connect, and turn traffic on, implementing the following steps.

1. Find or create the new traffic parameter row(s);

2. Associate the virtual links with the new traffic parameter row(s);

3. Activate the virtual links at the hosts and ISs by setting the atmVcpRowStatus to active;

4. Turn virtual link traffic on at ISs by setting the atmVcVrossConnectAdminStatus to up; and

5. Turn virtual link traffic on at hosts by setting the atmVclAdminStatus to up.

6.4.3.2 Topology Changes

A virtual connection topology change requires additional agent capabilities including hardware/software support. To accomplish a point to multipoint leaf addition, the following steps are required:

1. Define the virtual link to be added;
2. Define an additional row in the cross-connect table.

To accomplish a destination change, the following steps are required:

1. Delete the appropriate row in the cross-connect table;
2. Delete the appropriate virtual link (host and switch);
3. Define the virtual link to be added;
4. Define an additional row in the cross-connect table.

6.4.4   Tracing of VCs

In order to trace a VC through multiple switches and hosts, a manager needs to refer to information about how switches and hosts are interconnected. This includes both the location of neighboring switches and hosts and the topology of the links to neighboring switches and hosts. This is referred to as the neighbor information. The *location* of a switch or host, for our purposes, is the address of its SNMP agent. The topology of links between switches is captured in a mapping of each local interface to a neighbor switch or host and to an interface on that neighbor. In order to avoid topology ambiguity when switches have several parallel links, the interface names are needed on both sides of the link.

The neighbor information is expressed, for each physical interface on a switch, as the address of the agent and the interface name on the neighbor switch to which this interface is connected. It can be manually configured, or automatically discovered. In the latter case, an agent acquires the information by interchanging with other agents its address and its interfaces. The ATM Forum's ILMI neighbor discovery procedure can be used for such a purpose, as described in Chapter 5.

In the ILMI MIB, two objects are defined to support this auto-discovery, the MyIpNmAddress and the MyIfName, which is indexed by physical interface as discussed in Chapter 5. In order to represent neighboring information at the SNMP level, the AToM MIB also includes two objects, MyNeighborIpAddress and MyNeighborIfName, which are equivalent to the MyIpNmAddress and MyIfName in the Port Table of ILMI MIB. Both of them are in the interface configuration table, which is indexed by ifIndex. To illustrate this, consider the neighboring information for our example, which is maintained by the agents residing on host A, on the IS, and on host B.

To trace a virtual connection, the manager follows a simple algorithm involving the VPC/VCC cross-connect table and the neighbor information in the interface configuration table. To start, it identifies a starting point switch or host, a virtual connection, and a direction, probably by selecting an outgoing interface and VPI/VCI values. The outgoing VPI/VCI values equal the incoming VPI/VCI values on the next hop interface. The procedures to trace a virtual connection are as follows:

- From the entry in the interface configuration table, the manager determines the next hop switch (IP address) and the incoming interface name on that switch;

- From the VPC/VCC cross-connect table on the next hop switch, the manager determines the outgoing interface and the outgoing ifIndex/VPI/VCI values.

Table 6.13
Neighboring information for the example

System	Local ifIndex	My Neighbor's IP Address	My Neighbor's Interface Name
Host A	11	ipAddrIS	ifName13
IS	13	ipAddrA	ifName11
	3	ipAddrB	ifName1
Host B	1	ipAddrIS	ifName3

The algorithm is repeated until the VC terminates at a host. Since it has no cross-connections, a host does not support the VPC/VCC cross-connect table.

Let us start from host A, tracing the connection that we established. Table 6.13 contains neighboring information for the ATM devices involved. The connection is identified by the ifIndex/VPI/VCI, that is, 11/37/39.

1. Retrieve the neighboring information from the interface configuration table on host A.

   - get-request (atmInterfaceMyNeighborIpAddress.11);
   - response (atmInterfaceMyNeighborIpAddress.11 = ipAddrIS);
   - get-request (atmInterfaceMyNeighborIfName.11);
   - response (atmInterfaceMyNeighborIfName.11 = ifName11).

2. Determine the ifIndex value of incoming port of the IS. This can be obtained by looking up the ifTable of RFC 1573 or later version of the interface MIB on the IS using the atmInterfaceMyNeighborIfName (ifName11) from host A. The resulting value is 13.

3. Retrieve the cross-connect ID from the virtual link table. Because the outgoing VPI/VCI on the host A is the same as the incoming VPI/VCI of the IS, the VPI/VCI is 37/39. Thus, the identification for the incoming VCL is 13/37/39.

   - get-request (atmVclCrossConnectIdentifier.13.37.39);
   - response (atmVclCrossConnectIdentifier.13.37.39 = 3333).

4. Retrieve the identification of the outgoing virtual link from the VPI/VCI cross-connect table on the IS.

   - get-next-request (atmVcCrossConnectRowStatus.3333);
   - response (atmVcCrossConnectRowStatus.3333.3.25.27.13.37.39 = active(1)).

Thus the outgoing virtual link is 3/25/27. Note that there is only one entry in the table, as this is a point-to-point link. This would be indicated by the result from a further get-next-request.

5. Retrieve the neighboring information in the outgoing direction from the interface configuration table on IS.

   - get-request (atmInterfaceMyNeighborIpAddress.3);
   - response (atmInterfaceMyNeighborIpAddress.3 = ipAddrB);

- get-request (atmInterfaceMyNeighborIfName.3);
- response (atmInterfaceMyNeighborIfName.3 = ifName1).

6. Determine the ifIndex value of incoming port of the Host B. Again, this can be obtained by looking up the ifTable of RFC 1573 or a later version of the table on the host B using the atmInterfaceMyNeighborIfName (ifName1) from the IS. The resulting value is 1.

7. Retrieve the cross-connect ID from the virtual link table. Because the outgoing VPI/VCI on the IS is the same as the incoming VPI/VCI of the Host B, the VPI/VCI is 25/27. Thus, the identification for the incoming VCL is 1/25/27. Again, the manager issues:

```
get-request ( atmVclCrossConnectIdentifier.1.25.27 )
```

This would return "noSuchName" or "noSuchObject", indicating that the virtual link is terminated at this device.

Thus, the whole span of the virtual connection has been traced.

6.5 AToM MIB Versus ILMI MIB

AToM MIB and ILMI MIB share many factors. Both provide status and configuration information concerning the overall ATM interface, VPLs, and VCLs. Their messages follow SNMP syntax and encoding, as well as use SNMP protocol to transfer management information. They even have many overlapped managed object definitions. For that subset of management information that both MIBs have in common, AToM MIB has made every effort to retain identical semantics and syntax even though the MIB objects are identified using different OIDs. The interface management in AToM MIB is, in fact, based on the ILMI MIB defined in UNI 3.0/3.1.

However, there are some fundamental differences between these two MIBs. First, AToM MIB and the ILMI MIB have different roles in an ATM network. The ILMI MIB is intended for local interface management, although it can also be made available to network management (i.e., to an SNMP request from a NMS). For example, procedures such as address registration, auto-configuration of UNI version, support for topology discovery, as well as service registration in ILMI are defined only for local interface. This can also be seen from the way in which managed objects are chosen. The ILMI MIB only contains objects that are necessary for interface management. In contrast, the AToM MIB is designed for network management. In addition to the objects needed to manage ATM interface, it also includes those for network management such as VP/VC cross-connect tables.

Moreover, the ILMI has a different paradigm from the use of SNMP for network management, as depicted in Figure 6.8. In network management, an asymmetric *manager-agent* model is used in which a manager issues requests and an agent responds with responses and generates traps. In contrast, the ILMI protocol is used in a symmetric peer-to-peer fashion. An ATM IME can access, via the ILMI communication protocol, the ATM interface MIB information associated with its adjacent IME, acting as both a manager and an agent, It can, therefore, send requests/traps to the other side. In order for a network manager to access an ILMI MIB, a proxy (see Chapter 5) is normally needed.

The AToM and ILMI MIB also differ in the way SNMP messages are transmitted and utilized. The ILMI messages are encapsulated directly into AAL5 (or optionally AAL3/4) frames on the default/configured VCI/VPI, while the AToM management information is transmitted over most likely SNMP/UDP/IP/AAL5 although other protocols are possible. The ILMI messages are issued and consumed by an entity *local* to the ATM interface, whereas messages for the AToM MIB are issued and consumed by a management station.

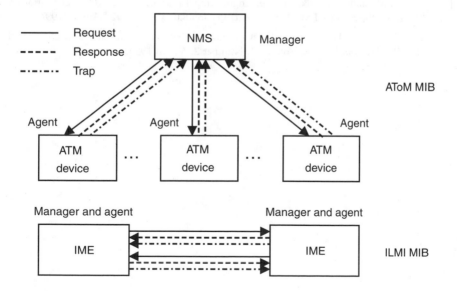

Figure 6.8 AToM versus ILMI paradigm.

In addition, the ILMI MIB and the AToM MIB conform to different versions of the standard. ILMI supports the SNMPv1, whereas AToM uses SNMPv2.

A summary is given in Table 6.14. A reasonable conclusion is that ILMI and AToM MIB tend to be complementary as they have different roles.

Table 6.14
Comparison of AToM MIB and ILMI MIB

	AToM MIB	ILMI MIB
Role	Network management	Local interface management
Management model	Manager-agent	Peer-to-peer
SNMP version	SNMPv2	SNMP v 1
Management information transmission	SNMP/UDP/IP/AAL5	SNMP/AAL5

References

[1]   Ahmed, M., and T. Tesink, Internet Engineering Task Force, RFC 1695, "Definitions of Managed Objects for ATM Management, Version 8.0 using SMIv2," August 1994.

[2]   Tesink, T., and T. Brunner, "(Re)Configuration of ATM Virtual Connections with SNMP," *The Simple Times,* Volume 3, Number 2, August 1994.

7

Physical Layer Management

ATM cells are transmitted over physical layer. ITU-T, the ATM Forum, and other regional standards organizations have been specifying a number of physical interfaces for ATM transmission. The transmission media for these interfaces include electrical, optical, and wireless, with transmission rates ranging from several megabits per second to several gigabits per second. The OAM&P of the underlying physical layer is crucial for the whole network to function properly. Hence, physical layer management is an integral part of the overall ATM network management. However, the management of the physical transmission system is different from that of a TCP/IP network.

Physical layer is medium-specific; that is, it depends on the physical transmission technology used. Three MIBs have already been defined for three different physical layer transmission lines:

- RFC 1407: "Definitions of Managed Objects for the DS3/E3 Interface Type" [1];
- RFC 1406: "Definitions of Managed Objects for the DS1 and E1 Interface Types" [2];
- RFC 1595: "Definitions of Managed Objects for the SONET/SDH Interface Type" [3].

This chapter details the SONET/SDH management as an example of transmission system management.

7.1   Introduction to Digital Transmission Systems

Plesiochronous digital hierarchy (PDH) and *synchronous digital hierarchy* (SDH)/ SONET are the most widely used technologies for ATM cell transmission. PDH was developed to carry the digitized voice high-speed transmission system more efficiently. It has evolved into the North American, European, and Japanese digital hierarchies where only a discrete set of fixed rates is available, namely, Nx 64 Kbps and then the next levels in the respective multiplex hierarchies as depicted in Figure 7.1. Plesiochronous means nearly synchronous. If two digital signals are plesiochronous, their transitions occur at *almost* the same rate, with any variation being constrained within tight limits.

In contrast, signals in SDH are synchronous, with the digital transitions occurring at exactly same rate. There may be, however, a phase difference between the transitions of the two signals; this would lie within specified limits. These phase differences may be due to propagation time delays or jitter

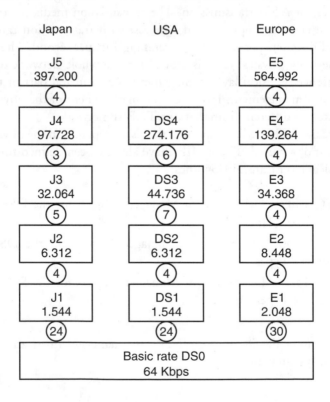

Figure 7.1 Plesiochronous digital hierarchy (megabits per second).

introduced into the transmission network. In a synchronous network, all the clocks are traceable to one primary reference clock. The SDH was proposed as a result of advancement of fiber optical transmission technology that enabled the transmission of large amounts of data over long distance at extremely high data rates.

Both SDH and PDH use a circuit switching approach where time is divided into time slots assigned to single channels during which users can transmit periodically. Basically, time slots denote allocated parts of the total available bandwidth. Before a data transfer commences, a connection between two end points must be established to allocate and reserve bandwidth for the entire duration. Usually this bandwidth is fixed and dedicated to a certain connection, even when no data is transmitted.

Although ATM stands for *asynchronous* transfer mode, it is actually synchronous in terms of its clock requirements. The ITU-T defines ATM as: "A transfer mode in which information is organized into cells; it is asynchronous in the sense that the recurrence of cells containing information from an individual user is not necessarily periodic." Thus, the actual meaning of *asynchronous* here is that the arrival of ATM cells does not have to fit into pre-allocated time slots as shown in Figure 7.2, which allows an ATM network to support advanced features like variable bit rates and variable bandwidth. Consequently, ATM is able to statistically multiplex the available bandwidth, resulting in a much higher average utilization of network resources.

7.2 SONET/SDH Physical Layer

SDH has been standardized by the ITU-T as an international standard for optical telecommunications transport. SONET is the U.S. counterpart of SDH defined by the *American National Standards Institute* (ANSI), which is responsible for developing U.S. industry standards. The comprehensive SONET/

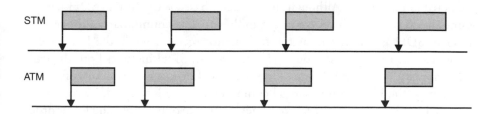

Figure 7.2 STM versus ATM.

SDH standard is expected to replace PDH as the transport infrastructure for worldwide telecommunications for the foreseeable future.

The increased configuration flexibility and bandwidth availability of SDH/SONET provides significant advantages over the older telecommunications transmission system. These advantages include the following:

- Definition of a transparent synchronous multiplexing synchronous structure for carrying lower level digital signals (such as E1, DS1, E3, and DS3), which means that a 64-Kbps channel can be accessed directly from the highest hierarchy, eliminating the need for back-to-back demultiplexing and multiplexing equipment used in PDH;

- Enhanced manageability due to the use of overhead bytes that permit management of the payload bytes on an individual basis and facilitate centralized fault sectionalization;

- Availability of a set of generic standards that enable products from different vendors to be interconnected;

- Definition of a flexible architecture capable of accommodating future applications, with a variety of transmission rates.

SONET and SDH are very close, but with just enough differences that they do not really interoperate. SONET is based on the STS-1 at 51.84 Mbps for efficient carrying of US DS3 (44.736-Mbps) signals, whereas SDH is based on the STM-1 at 155.52 Mbps for carrying ITU-T E4 (139.264-MHz) signals. As such, the ways in which payloads are mapped into these respective building blocks differ. Table 7.1 shows a comparison between SONET's *synchronous transport signal* (STS) and the SDH's *synchronous transport mode* (STM) signals.

7.2.1 Structure

The structure of a SONET/SDH network is divided into three levels as depicted in Figure 7.3. Although this model is based on SONET, SDH has the same structure but with different terminologies. The counterpart of the section level in SDH is referred to as the regenerator section, while the line level is known as the multiplex section level. The path in SDH includes both the path and tributary level in this model. However SDH does have two levels of virtual containers, the higher order virtual container and the lower order virtual container. The higher order virtual container is capable of multiplexing lower order virtual containers. Hence, the higher order virtual container level corresponds

Table 7.1
SDH and SONET Signals

U.S. SONET	ITU-T SDH	Bit Rate (Mbps)
STS-1	—	51.84
STS-3	STM-1	155.52
STS-12	STM-4	622.08
STS-24	STM-8	1244.16
STS-48	STM-16	2488.32
STS-192	STM-64	9953.28

to the path level, whereas the lower order virtual container level fits into the tributary level.

The roles for each level are as follows:

- *Section:* Spanning between two NEs capable of accessing, generating, and processing only regenerator *section overhead* (SOH);

- *Line:* Passing transparently through one or more regenerator sections terminated at NEs at each end that are capable of accessing, generating, and processing *line overhead* (LOH);

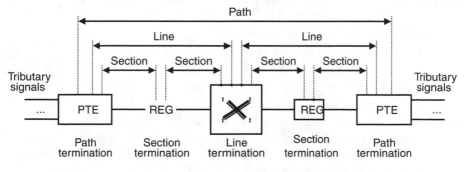

Figure 7.3 SONET/SDH levels.

- *Path:* Passing transparently through one or more line segments; extended between the point of assembly of the STS *synchronous payload envelope* (SPE) or the lower-level STM virtual container and its point of disassembly;

- *Tributary:* Similar to path, providing another level of assembly of low-data rate payloads. This level may not be present if the data rate of the payload is high (for example, when E4 is transmitted over STM-1).

SDH/SONET is widely used to transfer ATM cells as the physical layer of the latter. As introduced in Chapter 3, the physical layer in the B-ISDN is divided into three levels that correspond to SONET/SDH section, line, and path level, respectively.

7.2.2 Frame Format

The format of the SONET STS-1 frame is illustrated in Figure 7.4. The STS-1 frame is organized into a 90-column by nine-row structure of bytes. The frame duration is 125 μs (8000 frames per second). Hence, the STS-1 has a bit rate of:

$$9 \times 90 \text{ bytes/frame} \times 8 \text{ bits/byte} \times 8000 \text{ frames/} = 51.840 \text{ Mbps}$$

The first three columns of the STS-1 frame are for the transport over-head, which consists of a three-row (nine-byte) SOH for the section level and a six-row (18-byte) LOH for the line level. The remaining 87 columns constitute the STS-1 envelope capacity. The STS-1 payloads are encapsulated into the SPE, which occupies the envelope capacity.

The structure of STS-1 SPE is similar to that of the section level. Column 1 contains nine bytes, designated as the STS *path overhead* (POH). Two columns (columns 30 and 59) are not used for payload, but are designated as the *fixed stuff* columns. The 756 bytes in the remaining 84 columns are designated as the STS-1 payload capacity, which can be used to carry a T3 (44.736-Mbps) signal directly or further divided into virtual tributaries as listed in Table 7.2. The synchronous virtual tributaries (VTs) are used to transport lower speed signals. Again, VT contains VT POH and VT SPE.

SONET/SDH uses pointers to compensate for frequency and phase variations. Pointers allow the transparent transport of SPEs (either STS or VT) across plesiochronous boundaries. The use of pointers avoids the delays and loss of data associated with the use of large (125-microsecond frame) slip

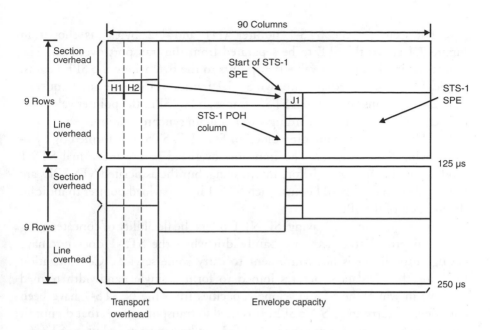

Figure 7.4 SONET frame structure.

Table 7.2
SONET Virtual Tributaries

Type	Bit Rate (Mbps)	VT Size	Max. Number of VTs per STS-1
VT 1.5	1.728	Nine rows, three columns	28
VT 2	2.304	Nine rows, four columns	21
VT 3	3.456	Nine rows, six columns	14
VT 6	6.912	Nine rows, 12 columns	7

buffers for synchronization. Pointers provide a simple means of dynamically and flexibly phase-aligning STS and VT payloads, thereby permitting ease of dropping, inserting, and cross-connecting these payloads in the network. Transmission signal wander and jitter can also be readily minimized with pointers.

For example, the STS-1 pointer (H1 and H2 bytes) as shown in Figure 7.4 allows the SPE to be separated from the transport overhead. The pointer is simply an offset value that points to the byte where the SPE begins. Typically, the SPE overlaps onto two STS-1 frames. If there are any frequency or phase variations between the STS-1 frame and its SPE, the pointer value will be increased or decreased accordingly to maintain synchronization.

STS-1 is the basic building block of SONET. STS-N is formed by byte-interleaving STS-1 frames. The transport overhead of the individual STS-1 modules are frame aligned before interleaving, but the associated STS SPEs are not required to be aligned because each STS-1 has a payload pointer to indicate the location of the SPE.

In addition to interleaving, SONET offers the flexibility of concatenating STS-1 to provide the necessary bandwidth when the STS-1 does not have enough capacity or is not convenient to carry some services. Concatenation links together N data structures joined to form a single bandwidth termed STS-Nc in which the STS envelope capacities from the N STS-1 have been combined to carry an STS-Nc SPE. It is used to transport signals that do not fit into an STS-1 payload. For instance, STS-1 can be concatenated up to STS-3c. Beyond STS-3, concatenation is done in multiples of STS-3c. Virtual tributaries can be concatenated up to VT-6 in increments of VT-1.5, VT-2, or VT-6.

SDH follows the same concept and frame structure, but the basic building block is the STM-1. From a formatting perspective, STS-3 is not identical to STM-1 even though the rate is the same. However, SONET STS-3c is the same as SDH STM-1, though there are other minor differences in overhead bytes (i.e., different places and slightly different functionality).

The transmission of ATM cells over STM-1 or STS-3c is depicted in Figure 7.5. Note that no VT level multiplexing is involved.

7.2.3 OAM&P Functions

SONET/SDH allows integrated network OAM&P in accordance with the philosophy of single-ended maintenance. In other words, one connection can reach all NEs (within a given architecture); separate links are not required for each NE. Remote provisioning provides centralized maintenance and reduced travel for maintenance personnel. Substantial overhead information is provided in SONET to allow quicker troubleshooting and detection of failures before they degrade to serious levels. Moreover, this information can be used to isolate faulty equipment and/or to switch hot-backup system automatically.

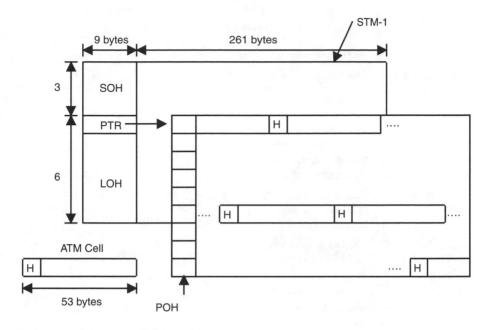

Figure 7.5 Encapsulation of ATM over STM-1.

Much of this overhead information is involved with alarm and in-service monitoring of the particular SONET/SDH levels. SONET/SDH alarms are defined as follows:

- *Anomaly:* The smallest discrepancy that can be observed between the actual and desired characteristics of an item. The occurrence of a single anomaly does not constitute an interruption in the ability to perform a required function.

- *Defect:* The density of anomalies has reached a level where the ability to perform a required function has been interrupted. Defects are used as input for performance monitoring, the control of consequent actions, and the determination of fault cause.

- *Failure:* The inability of a function to perform a required action persisted beyond the maximum time defined.

Table 7.3 summarizes SONET/SDH OAM&P signals.

Table 7.3
SONET/SDH OAM&P Signals

Signal Name	Description	Level			
		Section	Line	Path	VT
Loss of signal (LOS)	The amplitude of the relevant is dropped below prescribed limits for a prescribed period	√			
Out of frame (OOF)	The position of the frame AL bytes in the incoming bit stream is unknown	√			
Loss of frame (LOF)	LOF state occurs when the OOF state exists for a specified period of time	√			
Loss of pointer (LOP)	The LOP state is one resulting from a defined number of consecutive occurrences of certain conditions that are deemed to have caused the value of the pointer to be unknown			√	√
AIS	An alarm sent downstream indicating that an upstream failure has occurred		√	√	√
Remote error indication (REI)	An indication returned to a transmitting node (source) that an errored block has been detected at the receiving node (sink); this indication was also known as *far end block error* (FEBE)		√	√	√
Remote defect indication (RDI)	A signal returned to the transmitting terminating equipment upon detecting a LOS, LOF, or AIS defect; RDI was previously known as FERF		√	√	√

Signal Name	Description	Level			
Remote failure indication (RFI)	A defect that persists beyond the maximum time allocated to the transmission system protection mechanisms; when this situation occurs, an RFI is sent to the far end and will initiate a protection switch if this function has been enabled		√	√	√
Code violation (CV)	*Bit interleaved parity* (BIP) errors that are detected in the incoming signal; CV counters are incremented for each BIP error detected	√	√	√	√

7.3 RFC 1595 SONET/SDH MIB

As mentioned in Chapter 6, the RFC 1595 SONET/SDH MIB was also developed by the IETF AToM MIB Working Group. This MIB and RFC 1406 and RFC 1407 are collectively referred to as trunk MIBs, since E1/DS1, E3/DS3, and SONET/SDH transmission systems are deployed extensively as trunks in telecommunications. These MIBs are very similar to one another in terms of the structure and even managed object definitions. The current status of the SONET/SDH MIB is *proposed standard.* SONET/SDH MIB conforms to the SNMPv2 framework.

7.3.1 Management Model

The trunk MIBs differ from ordinary Internet MIBs in that they define managed objects in a totally different way, which, in turn, reflects the differences between traditional telecommunications and data communications in managing their respective networks.

In ordinary Internet MIBs, performance statistics are collected by counting various types of events, such as correct and errored input/output packets or bytes derived from the packets. All the counters are continuous and increase monotonically. An SNMP network manager relies on calculating the difference between subsequent samples to obtain the current performance of the managed system. This is a natural choice for TCP/IP networks since they are packet-switched where packet errors can be easily counted.

In telecommunications, however, the focus is on the provisioning of circuit-switched services. Reliability and availability are of higher priority. Accordingly, statistics collected should reflect such requirements. Consequently, the performance of a transmission system is evaluated in terms of a set of parameters such as *errored seconds* (ESs) and *severely errored seconds* (SESs). Traditional practice in telecommunications is to keep performance information in 15-minute intervals. Trunk MIBs defined managed objects accordingly.

Resources in SONET/SDH interfaces are managed by the MIB II Interfaces Group and the SONET/SDH MIB as a medium-specific extension. The management is carried out at five levels, as depicted in Figure 7.6. The SONET/SDH entities are modeled as below, and the exact configuration and multiplexing of the levels is maintained in the ifStackTable (see Chapter 2 for more details).

The usage of ifEntry to manage different SONET/SDH levels are as follows:

- One ifEntry for the combined SONET physical, section, and line levels for a particular port.

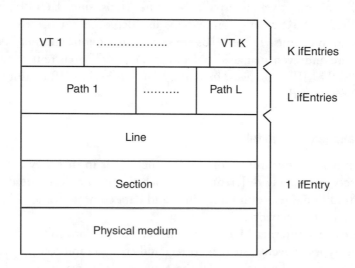

Figure 7.6　SONET/SDH management levels (RFC 1595).

- One ifEntry for each SONET path; for example, an STS-3 has three entries, whereas STS-3c has only one entry in the ifTable, though the transmission rates for both systems are the same.
- One ifEntry for each SONET VT; again, an STS-1 that supports 28 VT 1.5 has 28 ifEntries.

Note that the management of ATM networks typically does not involve the VT level SONET/SDH management as ATM cells are encapsulated into the path level directly. Instead, the TC group in the AToM MIB is used to monitor the performance of the TC from SONET/SDH to ATM layer. The ifEntries that are used to transfer ATM cells over STS-3c or STM-1 are shown in Figure 7.7.

7.3.2 MIB Structure

The structure of the SONET/SDH MIB is shown in Figure 7.8, which consists of eight groups. The medium group contains configuration information such as interface type and line code for the PM level of SONET/SDH ports. The section group comprises two subgroups for the section level, the section current

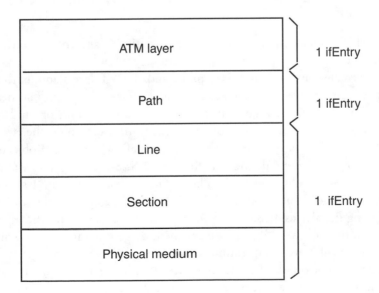

Figure 7.7 Management levels for ATM over SONET/SDH.

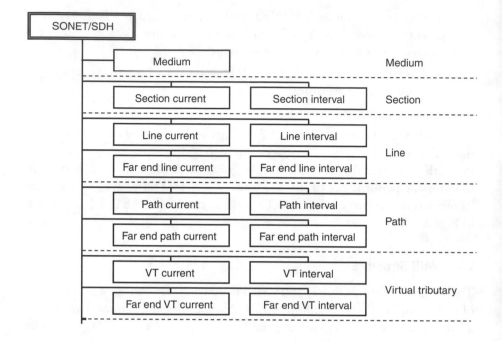

Figure 7.8 SONET/SDH MIB structure.

group, which holds current status and error statistics, and the section interval group, which holds a history of error statistics for every 15 minutes.

The management of the line, path, and the VT levels adopts the same model, which involves a near-end group and a far-end group. The near-end group, almost identical to the section group, contains local history and current information. The far-end group contains similar information for the far-end.

Error statistics are the major concern in defining managed objects. The most important objects defined include ESs, SESs, *severely errored framing seconds* (SEFSs), UASs, and CVs. These parameters, except the last one, are derived from the alarms listed in Table 7.3. RFC 1595 details the mechanism of mapping the alarms defined in SONET/SDH standards into these parameters. Note that all these objects are specified as gauges rather than counters, because they will never wrap round. Consequently, SNMP managers monitor the values of these objects directly. In contrast, only the delta values make sense for counters. The delta values are calculated from the base value and values obtained from subsequent queries.

7.3.3 Medium Group

This group contains general configuration information for SONET/SDH ports. The medium table is indexed by the ifIndex that is also used to index the entry for this SONET/SDH interface in the MIB II ifTable. Table 7.4 provides a detailed description of defined managed objects.

Table 7.4
Objects for the Medium Group

Prefix	sonetMedium		
Category	Name	Definition	Comments
Configuration	Type	Indicates whether this interface is SONET or SDH	
	TimeElapsed	The number of seconds that have elapsed since the beginning of the current error-measurement period	
	ValidIntervals	The number of previous intervals for which valid data has been stored	$4 \leq N \leq 96$ default $= 32$
	LineCoding	Line coding for this interface, such as B3ZS and CMI for electrical signals, and *non-return to zero* (NRZ) and return to zero are used for optical signals	
	LineType	Line type for this interface indicating the type of the underlying transmission medium	
	CircuitIdentifier	Transmission vendor's circuit identifier	

7.3.4 Section Management

The management at the section level involves two tables. The SONET/SDH section current table contains various statistics being collected for the current 15-minute interval. The section interval table, on the other hand, keeps a history of statistics gathered over a maximum of the previous 24 hours of operation that is broken into up to 96 completed 15-minute intervals. A system is required to store at least four completed 15-minute intervals. The default value is 32 intervals. This value is configurable through the SONET/SDH medium group.

     The objects defined in the section current table and the section interval table are almost identical except that the former contains an object to indicate current status of the interface at section level. The latter, on the other hand, provides an object identifying the interval for which the set of statistics is available. As a result, the current table is indexed by the related ifIndex only, while both the ifIndex and an interval number identify an entry in the interval table. Tables 7.5 and 7.6 list the objects defined for the current and the interval groups, respectively.

Table 7.5

Objects for the Section Current Group

Prefix	sonetSectionCurrent		
Category	Name	Definition	Comments
Status	Status	A bit map represents the status of this interface as follows:	
		Bit Position Description	
		1 sonetSectionNoDefect	
		2 sonetSectionLOS	
		4 sonetSectionLOF	
Error Statistics	Ess	Number of Ess	Encountered by a SONET/SDH section in the current 15-minute interval
	SESs	Number of SESs	
	SEFSs	Number of SEFSs	
	CVs	Number of CVs	

Table 7.6
Objects for the Section Interval Group

Prefix	sonetSectionInterval		
Category	Name	Definition	Comments
Index	Number	Identifies the interval for which the set of statistics is available; the interval identified by 1 is the most recently completed 15-minute interval, and the interval identified by N is the interval immediately preceding the one identified by N - 1	
Error Statistics	Ess	Number of Ess	Encountered by a SONET/SDH section in the current 15-minute interval
	SESs	Number of SESs	
	SEFSs	Number of SEFSs	
	CVs	Number of CVs	

7.3.5 Line/Path/VT Management

The management of SONET/SDH at the line, path, and VT levels is almost the same: Each involves two groups defined in the SONET/SDH MIB (i.e., a near-end group and a far-end group), reflecting the fact that the line/path/VT level is end-to-end. The near-end group consists of current and interval tables collecting statistics based on the near-end alarms. This group is almost identical to the section group in terms of objects defined. The major difference is that an object that collects the number of UASs replaces the object denoting the number of SEFSs for the section level. In addition, there are some minor differences in the definition of the status objects at different levels.

Similarly, a far-end group consists of far-end current and far-end interval tables collecting the same error statistics as the corresponding near-end group based on the alarms sending from the far-end, particularly the FEBE signal or RDI signal. This group may only be implemented by SONET/SDH systems that provide FEBE information. Note that no status object is defined in the far-end current table, as no FEBE information will be available if the link is in error.

Last, an object to indicate the width of the signal is added to the path and VT current groups, respectively. The purpose of this object is to explicitly identify the type of the path/VT, although the transmission rate can be obtained from the ifTable.

References

[1]   Cox, T., and K. Tesink, Internet Engineering Task Force, RFC 1407, "Definitions of Managed Objects for the DS3/E3 Interface Type," January 1993.

[2]   Baker, F., and J. Watt, Internet Engineering Task Force, RFC 1406, "Definitions of Managed Objects for the DS1 and E1 Interface Types," January 1993.

[3]   Brown, T., and K. Tesink, Internet Engineering Task Force, RFC 1595, "Definitions of Managed Objects for the SONET/SDH Interface Type," March 1994.

[4]   ITU-T Recommendation G.707, "Synchronous Digital Hierarchy Bit Rates," March 1993.

[5]   ITU-T Recommendation G.708, "Network Node Interface for the Synchronous Digital Hierarchy," March 1993.

[6]   ITU-T Recommendation G.709, "Synchronous Multiplexing Structure," March 1993.

[7]   ITU-T Recommendation G.783, "Characteristics of Synchronous Digital Hierarchy (SDH) Multiplexing Equipment Functional Blocks," January 1994.

8

Private Network-Network Interface

8.1 Introduction

PNNI is the protocol that enables the building of multivendor, interoperable ATM switching networks. The abbreviation PNNI is intentionally defined to mean either *private network node interface* or *private network-to-network interface*. The reason is that the PNNI protocol is designed to be equally applicable and efficient whether operating between two switches as a network-node interface or between two groups of switches as a network-network interface. In fact, this similarity between the two cases is the key to how the protocols are able to elegantly scale up to handle extremely large networks.

A private ATM network differs from a public one in that it uses NSAP format ATM addresses defined by the ATM Forum within the network. Public networks that use E.164 numbers for addressing will be interconnected using a different NNI protocol stack based upon the ITU-T B-ISUP signaling protocol and the ITU-T MTP level 3 routing protocol.

PNNI is a very complex protocol. This complexity arises from two goals of the protocol: to allow for much greater scalability than would be possible with any existing protocol and to support true QoS-based routing.

It is envisaged that the so-called private network may range from a small number of switches to very large private ATM networks composed of several corporate ATM networks interconnected at multiple points. In other words, PNNI should be equally applicable between switching systems in a multivendor ATM LAN and between switching systems in a multivendor global ATM network. To scale such networks, PNNI adopts a recursive hierarchical model that fully exploits the unprecedented level of hierarchical structure of the ATM address to support greater degrees of scalability within ATM networks than is possible within any other network.

With regard to QoS-based routing, the PNNI protocols are designed to support the set-up of ATM connections with guaranteed QoS. Essentially, all aspects of the intended traffic flow and desired QoS, from bandwidth and burstiness to latency and jitter, can be specified and used as the basis of call routing. Specifically, the PNNI routing protocols allow ATM switches to exchange between themselves not only reachability information but also QoS metrics such as bandwidth, guaranteed cell delays, or jitter that a particular switch can make available to a new ATM connection. The reachability information is to let all switches know which ATM addresses are reachable through any other switch.

PNNI is very sophisticated and flexible. It standardizes switch-to-switch signaling and topology information distribution within the ATM network. PNNI also allows the connection management systems of various vendors to communicate and provide SVC routing across many ATM systems. Besides, it supports all UNI 3.1 and some UNI 4.0 capabilities, such as anycast, soft PVPC/ PVCC, as well as physical and virtual links tunneling over VPCs. In addition, it is capable of working in the presence of a partitioned area and interoperating with external routing domains (not necessarily using PNNI).

While the "p" in PNNI stands for private, many carriers that are looking to reduce their OAM&P costs are applying PNNI to their public networks. It is less expensive to implement and administer, and with its hierarchical routing protocol, it scales to very large networks. For these reasons, PNNI will be very important for many service provider organizations, such as those delivering data overlay services, which are looking to leverage the advantages of internetworking within their own multiservice networks.

The current version of PNNI protocol [1–4] was developed by the ATM Forum PNNI Working Group. It consists of two key protocols: the PNNI routing protocol and the PNNI signaling protocol. The former is a hierarchical, dynamic topological-state routing protocol for determining reachability, while the latter defines signaling method for propagating user calls. These protocols will be introduced in the Section 8.2.

The key feature of the PNNI hierarchy mechanism is its ability to automatically configure itself in networks. As a result, PNNI defines a very elaborated MIB [4] for this purpose. In addition, it defines a MIB [2] to manage soft PVPC/PVCC. The management framework and the PNNI MIB will be introduced in Section 8.3.

8.2 PNNI Protocols

8.2.1 Routing

PNNI routing protocol is designed to route the virtual circuit connection request through the ATM network. The routing of the connection request, and hence of any subsequent data flow, is governed by the PNNI routing protocol based upon the destination address and the traffic and QoS parameters requested by the source end system. Thus, the NNI protocols are to ATM networks what routing protocols are to current routed network.

It is interesting to note that the operation of routing a signaling request through an ATM network—somewhat paradoxically, given ATM's connection-oriented nature—is superficially similar to that of routing connectionless packets within existing network layer protocols such as IP. This is due to the fact that prior to connection set-up, there is, of course, no connection for the signaling request to follow. As such, a VC routing protocol can use some of the concepts underlying many of the connectionless routing protocols that have been developed over the last few years [5].

The overall architecture of the PNNI routing protocol is based on research and development work done in the IETF, particularly at the Nimrod Working Group [6]. The Nimrod activity has been focused on routing and addressing architecture for TCP/IP. The PNNI routing protocol partakes of the philosophy and network view of Nimrod. However, it is a distinct protocol evolving from the Nimrod routing protocol.

8.2.1.1 Identification of PNNI Entities

The PNNI protocol is designed to be applied both to small networks of a few switches and to a possible future global ATM Internet comprising millions of switches. Such scalability is well beyond that of any single routing protocol today. For example, the Internet routing protocols only support two levels of networks: intradomain routing protocols that scale to large enterprise networks and interdomain protocols that interconnect such lower level networks. The aim of the PNNI is, however, to build a single protocol that could perform at all levels within a network.

The key to such a scalable protocol is hierarchical network organization, with summarization of reachability information between levels in the hierarchy. To support this hierarchy, the PNNI model defines a uniform network model at each level of the hierarchy by utilizing the ATM Forum's private model. Note that PNNI routing operates only on the first 19 octets of the AESA.

/se page 99

8.2.1.1.1 Peer Group

PNNI is a hierarchical routing protocol that organizes switching systems into *peer groups* (PGs), logical collections that contain network nodes at a given level. A PG is identified by a *peer group identifier* (PGI) encoded using 14 octets, which is essentially the prefix of the prefix of the PG. Figure 8.1 depicts the structure of the PGI.

PGI consists of two fields:

- Octet 1: Level indicator indicates the position in the PNNI hierarchy at which a particular node or PG exists and ranges from 0 to 104;
- Octet 2–14: ATM address prefix (i.e., the IDP and HO-DSP parts of an ATM address).

The level indicator corresponds to the number of bits information field. The value of the identifier information must be encoded with the $104 - n$ right-most bits set to zero, where n is the value of level indicator. As an example, the PGI of node A1 in Figure 8.4 is as below, where the level is 96 (0x60).

0x60.47.0091.8100.0011.0005.0103.0400

8.2.1.1.2 Logical Node

Each level in the hierarchy consists of a set of logical nodes, interconnected by logical links. Each logical node is assigned a unique node identifier encoded as 22 octets consisting of a one octet level indicator and a 21-octet opaque value—i.e., these 21 octets have no implied internal structure. Hence, a node receiving a node identifier in a PNNI packet must make no assumptions about the internal structure. There are two types of logical nodes, physical nodes and logical group nodes.

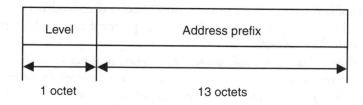

Figure 8.1 Structure of PGI.

Physical node. At the lowest level, each logical node represents a physical switching system consisting of a single physical switch, or a network of switches that internally operate a proprietary NNI protocol and support the PNNI protocol for external connectivity. At this lowest level, each switching system must be assigned a unique AESA. This address is used for establishment of SVCCs used in PNNI. Figure 8.2 illustrates the structure of the node identifier for physical nodes, which consists of three fields as follows:

- Octet 1: Level indicator specifies the level of the node's containing PG;

- Octet 2: 160; this is to indicate that this is not a PG;

- Octet 3–22: AESA, as described previously.

The physical node identifier of node A1 in Figure 8.4 is as below. The second octet is 160 (0xA0), which identifies it as a physical node.

0x60.A0.47.0091.8100.0011.0005.0103.0400. 00603E5ADB01.00

Logical group node. PGs are organized hierarchically and are associated with a higher level PG. Within its parent PG, each PG is represented as a single logical node. A logical group node is also identified by a node identifier, but with a slightly different format from the physical one, which is illustrated in Figure 8.3. Logical group identifier consists of four fields as follows:

- Octet 1: Level indicator specifies the level of the PG containing the logical group node;

- Octet 2–15: PG Identifier as described above;

- Octet 16–21: ESI as described above;

- Octet 22: Zero.

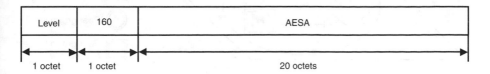

Figure 8.2 Structure of physical node identifier.

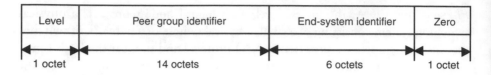

Figure 8.3 Structure of logical node identifier.

The logical node identifier for logical node A in the Figure 8.4 is as below, where the level indicator is 72 (0x48).

0x48.60.A0.47.0091.8100.0011.0005.0103.0400. 00603E5ADB01.00

8.2.1.1.3 Logical Link

A logical link is an abstract representation of the connection between two logical nodes. This includes physical links, individual VPCs, and parallel physical links and/or VPCs. A logical link is identified by the logical node identifier of either node at the end of that link and the port identifier assigned by that node. The port identifier is a 32-bit number assigned by a node to unambiguously identify a point of attachment of a logical link to that node. Port identifiers are only meaningful in the context of the assigning node, identified by its node

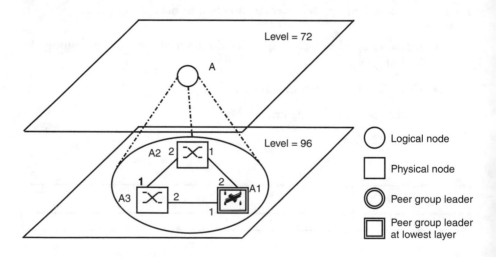

Figure 8.4 Example of a logical group.

identifier. For instance, the link between switching system A1 and A2 in Figure 8.4 is connected from port 2 of A1 to port 1 of A2. The link can only be identified by the logical node identifier and the port identifier.

In brief, the model described above is recursive in that the same mechanisms used at one level are also used at the next level. At higher levels, the default for a PG ID is a prefix on a lower level PG ID. The PG ID of a parent must be shorter than the prefix of its child PG ID; this makes it easy to determine the relationship between two PGs and precludes the formation of a PG hierarchy loop. Hence, the PG ID becomes smaller as the hierarchical level becomes higher. By using the prefix of AESA to identify levels in the network hierarchy, the actual number of levels that can be supported is almost limitless. The maximum levels are 104, although no more than a half dozen or so will likely ever need to be used, and even then only within the very large global networks.

8.2.1.2 Routing Protocol

PNNI is a hierarchical, dynamic link (topological) state routing protocol, in which all nodes within a given level maintain a complete *map* of the network and perform local computation of best routes based on this internal map. The PNNI routing protocol performs the functions detailed in Sections 8.2.1.2.1–8.2.1.2.5.

8.2.1.2.1 Discovery of Neighbors and Link Status (Hello)

When a logical link becomes operational, the attached nodes initiate an exchange of information by Hello protocol via a VCC used as a PNNI *routing control channel* (RCC). This is either a well-known VCC (VPI=0, VCI=18) if nodes are in the lowest level, or an established SVCC when they are higher level nodes. Hello packets are sent periodically by each node on this link specifying the AESA, node ID, PG ID, and its port ID for the link. In this way, the hello protocol makes the two neighboring nodes known to each other. The hello protocol runs as long as the link is operational. It can therefore act as a link failure detector when other mechanisms, such as the OAM continuity check cells, fail.

The PG IDs are exchanged between neighboring nodes in hello packets. These IDs are used to determine whether the nodes belong to the same PG. If they have the same PG ID then they belong to the same PG. The link between the nodes that are in the same PG is termed as the horizontal, or inside, link. If the exchanged PG IDs are different, then the nodes belong to different PGs. The link between lowest level outside nodes is called outside link.

Consequently, the nodes are called border nodes. An uplink is the connection from a border node to an upnode. There is another possibility; that is, the link connects a PNNI node and a node that does not support PNNI protocol. In this case, the link is called exterior link. Figure 8.5 depicts different logical links.

8.2.1.2.2 Information Exchange Within a Peer Group

Within a PG, all nodes obtain the identical topological database and exchange full link state information with each other. Such information is passed within *PNNI topology state packets* (PTSP), which contain various *type-length-value* (TLV)-encoded *PNNI topology state elements* (PTSE).

PTSEs represent the smallest collection of PNNI routing information that is flooded as a unit among all logical nodes within a PG. A node's topology database consists of a collection of all PTSEs received, which represent that node's present view of the PNNI routing domain. In particular, the topology database provides all the information required for computing a route from the given node to any address reachable in or throughout that domain.

The topology information consists of three parts as follows:

1. Nodal information: Describes node's own identity and capabilities, information used to elect the *PG leader* (PGL), as well as information used in establishing the PNNI hierarchy.

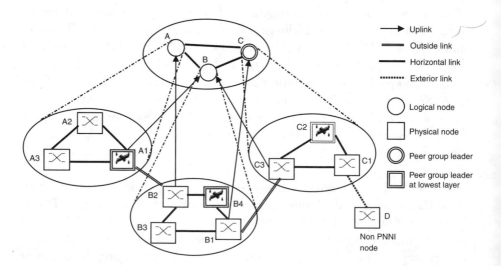

Figure 8.5 PNNI logical links.

2. Topology state information: Consists of link state parameters that describe the characteristics of logical links and nodal state parameters that describe the characteristics of nodes. This will be discussed further in Section 8.2.2.2.

3. Reachability information: Consists of addresses and address prefixes that describe the destinations to which calls may be routed. This information is advertised in PTSEs by nodes in the PNNI routing domain. Note that the scope of advertisement of the group addresses is a function of how the network administrator maps the administrative scope of a registered node to the corresponding PNNI hierarchy.

When neighboring nodes find out that they belong to the same PG, they proceed to synchronize their topology databases. This process results in the two nodes having identical topology databases.

Once the nodes have synchronized their databases, they flood PTSPs throughout the PG (i.e. across horizontal links) to ensure that each node in a PG maintains an identical database. Flooding is the reliable hop-by-hop propagation of PTSEs throughout a PG. It is the advertising mechanism in PNNI. Flooding is an ongoing activity; that is, each node issues PTSPs with PTSEs that contain updated information. PTSEs are reissued both periodically and on an event-driven basis.

8.2.1.2.3 Vertical Information Exchange

A PG is represented in the next hierarchical level by a single node called a *logical group node* (LGN). The PGL node performs the functions needed to fulfill this role.

PGL is selected through an election mechanism and is based upon a leadership priority and the switches' node ID. Each PGL is identified by a unique ATM address; if a node acts as a PGL within multiple levels of PGs, then it must have a unique ATM address at each of those levels.

PGLs within each PG have the responsibility of aggregating and exchanging PTSPs with their peer nodes within the parent PG to inform those nodes of the child group's reachability and topology aggregation information. Reachability refers to summarized address information needed to determine which address can be reached through the lower level PG. Topology aggregation refers to the summarized topology information needed to route into and across this PG. Note that there is a filtering function inherent in the summarization process that propagates only the information needed by the higher layer. PTSEs never flow up hierarchy. Only summarized information is advertised within PTSEs originated by the logical node and flooded to its peers.

Similarly, recursive information obtained by the PGL about the parent group and that group's parent groups are then fed down by the PGL into the child group. The child nodes can then obtain knowledge about the full network hierarchy in order to construct full source routes. The information consists of all PTSEs it originates or receives via flooding from other members of the LGN's PG. Each PTSE that flows down to a PGL is flooded across that PG. This gives every node in a PG a view of the higher levels into which it is being aggregated. Figure 8.6 depicts the vertical and horizontal information flows.

8.2.1.2.4 Information Exchange Across Border

When neighboring nodes conclude from the hello protocol that they belong to different PGs, they become border nodes. Links between border nodes in different PGs are called outside links. There is no database exchange across outside links.

Border nodes extend the hello protocol across outside links to include a nodal hierarchy list about their respective higher level PGs and logical group nodes representing them in these PGs. This information allows the border nodes to determine the lowest level PG common to both border nodes.

The mechanisms described above allow each node to know the complete topology, including nodes and links within its PG, as well as the complete summarized topology of the higher level parent PG, grand-parent PG, etc. In order for the node to realize which border nodes have connectivity to which higher level nodes, the border nodes advertise the links as up-links to those higher leve lnodes.

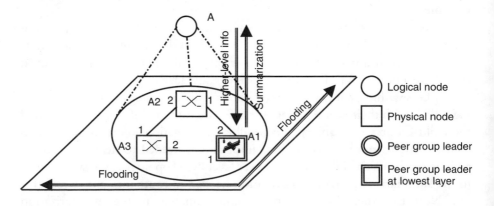

Figure 8.6 Information flows in PNNI.

8.2.1.2.5 Topology Aggregation

Topology aggregation is the notion of reducing nodal as well as link information to achieve scaling in a large network. It is necessary not only to reduce the complexity but also to hide the topology internals of PGs in the interest of security. Topology aggregation includes link aggregation and nodal aggregation.

Link aggregation refers to the representation of some set of links between the same two PGs by a single logical link. This is the responsibility of logical group nodes. A logical group node examines all of the up-link advertisements from its child PG to a specific up-node. All up-links to the same up-node with the same aggregation token as the result of configuration are aggregated into a single link. The resulting link could be either a horizontal link, if the up-node is a peer of the LGN, or an induced up-link otherwise.

Nodal aggregation is the process of representing a child PG by a LGN in its parent PG. PNNI uses the complex node representation technique, which allows a PG to be modeled at higher levels, for advertising purposes, not as a single node but as a *complex node,* with an internal structure. This technique improves the accuracy of nodal aggregation.

Figure 8.7 illustrates the recursive representation of an ATM network using the PNNI model, which consists of three hierarchical levels.

8.2.2 PNNI Signaling

ATM signaling uses the *one-pass* method of connection set-up, which is the model used in all common telecommunications networks, such as the telephone network. That is, a connection request from the source end system is propagated through the network, setting up the connection as it goes, until it reaches the final destination end system. This process is depicted in Figure 8.8. PNNI signaling protocol is used to establish point-to-point and point-to-multipoint connections across ATM private network. This is based on the UNI signaling specification, with additional provisions added to support source routing, crankback, and alternate routing in case of a failure in the connection set-up. PNNI also includes the support of soft permanent VPCs/VCCs.

Specifically, PNNI is based on a subset of UNI 4.0 signaling. It does not support some UNI 4.0 features such as proxy signaling, leaf-initiated join capability, or user-to-user supplementary service, but adds new features that pertain to the use of PNNI routing for dynamic call set-up. In addition, PNNI signaling differs from UNI 4.0 signaling in that it is symmetric. The reason for this is

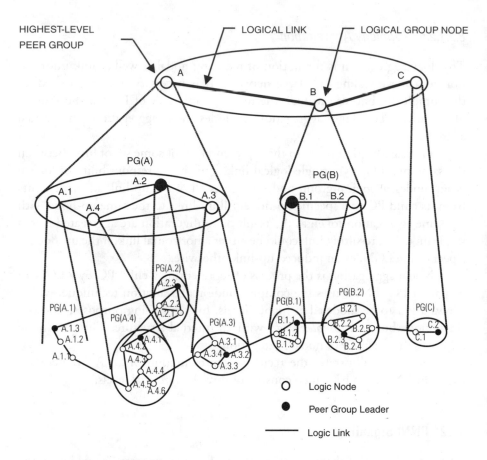

Figure 8.7 PNNI network hierarchy model. (*Copyright 1996The ATM Forum.*)

that the PNNI signaling works via symmetrical NNI nodes rather than across asymmetric UNI in UNI signaling protocols.

8.2.2.1 Source Routing

ATM is connection-oriented, which means that a path selected by PNNI for the establishment of a virtual connection will remain in use for a potentially extended period of time. In order to ensure a loop free path selection and user-specified QoS and to allow sources to utilize local policy, the source routing technique is chosen for PNNI.

The originating switch examines the full set of maps, as collected by the routing protocol, for the local PG and all visible higher level PGs, including the

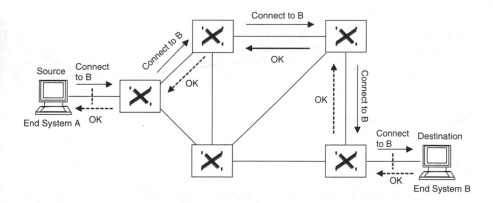

Figure 8.8 Connection set-up.

map for the PG which advertises reachability to the called party address. Using this, it then builds a sequence of paths. At the top is a path across the highest necessary level of PGs to get to the destination from a parent that contains the source. Then, for each intermediate PG level, a source route is built across that PG. The result is a stack of source routes across successively higher levels of PG. Each source route is known as a *designated transit list* (DTL).

As a consequence, the low-level switches need a lot of topology information. However, it fully prevents loops and allows local switches to make path choices based on local policy and to know that those choices will be obeyed. Additionally, the source-routed behavior of the signaling removes one of the significant design difficulties of routing protocols. That is, it is no longer necessary for all switches to have the same routing information.

Note that the source routing in PNNI is, in fact, partial source routing in that the source route computed by the source switch is only hierarchically complete, rather than a truly complete source route. Figure 8.9 is an example of the global topology as seen by node A.1.2. The node does not have detailed information of the internal structure of the PGs other than those in which it participates. As a result, whenever a call enters a PG from the outside, the entry switch must create a source route (DTL) to transit the PG.

In order to support source routing, PNNI signaling protocol incorporates additional *information elements* (IEs) for such NNI-related parameters as DTLs. PNNI signaling is carried across NNI links on the same VC (VPI=0, VCI=5), which is used for signaling across the UNI. The VPI value depends on whether the NNI link is physical or virtual.

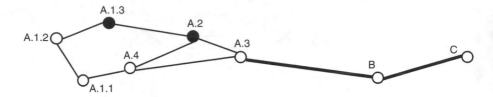

Figure 8.9 Global topology as seen by node A.1.2.

8.2.2.2　QoS Support

One of the essential requirements of ATM is its support for guaranteed QoS in connections. Hence, a node requesting a connection set-up can request a certain QoS from the network and be assured that the network will deliver that QoS for the life of the connection. To deliver such QoS guarantees, ATM switches implement a function known as *connection admission control* (CAC). Whenever a connection request is received, the switch performs the CAC function. That is, the switch determines whether setting up the connection violates the QoS guarantees of established connections based on the traffic and QoS parameters of the requested connection. The switch accepts the connection only if violations of current guarantees are not reported. CAC is a local switch function and is dependent on the architecture of the switch and local decisions on the strictness of QoS guarantees [5].

The VC routing protocol must ensure that a connection request is routed along a path that leads to the destination and has a high probability of meeting the QoS requested in the connection set up, that is, of traversing switches whose local CAC will not reject the call. To calculate the source route, the source PNNI node needs not only the reachability information but also the information on resource availability within the network—i.e., the topology state information. All the topology state parameters defined by the PNNI Phase 1 are link state parameters except the *restricted transit flag* parameter, which is used for policy control.

There are two types of topological state parameters: nonadditive topological state attributes used to determine whether a given network link or node can meet a requested QoS and additive topological metrics that are used to determine whether a given path can meet the requested QoS. The additive topological metrics consists of a set of concatenated links and nodes (with summarized link metrics).

The current sets of topology metrics (additive) are the following.

- *Maximum cell transfer delay* (MCTD) per traffic class;

- *Maximum cell delay variation* (MCDV) per traffic class;

- Maximum *cell loss ratio* for CLP = 0 cells, (CLR0) for the CBR and VBR traffic classes;

- Maximum cell loss ratio for CLP = 0 + 1 cells, (CLR0 + 1) for the CBR and VBR traffic classes;

- Administrative weight, a value set by the network administrator, that is used to indicate the desirability or otherwise of a network link.

The current sets of topology attributes (nonadditive) are the following.

- *Maximum cell rate* (maxCR): Maximum bandwidth in cells per second to the specified traffic class;

- *Available cell rate* (ACR): Available bandwidth in cells per second, per traffic class;

- *Cell rate margin* (CRM): Difference between the effective bandwidth allocation per traffic class and the allocation for sustainable cell rate; this is a measure of the safety margin allocated above the aggregate sustained rate;

- *Variance factor* (VF): A relative measure of CRM margin normalized by the variance of the aggregate cell rate on the link.

The topology state information is encoded in TLV format within PTSEs. The representation of attributes of nodes and links has been carefully designed to be extensible to support new traffic classes and new metrics in the future.

Note that CAC is a local matter, which means that the CAC algorithm performed by any given node is both system-dependent and open to vendor differentiation.

The PNNI protocol tackles this problem by defining a *generic CAC* (GCAC) algorithm. This is a standard function that any node can use to calculate the expected CAC behavior of another node, given that node's advertised additive link metrics and the requested QoS of the new connection request. The GCAC is an algorithm that was chosen to provide a good prediction of a typical node-specific CAC algorithm, while requiring a minimum number of link state metrics. Individual nodes can control the degree of stringency of the GCAC calculation involving the particular node by controlling the degree of laxity or conservatism in the metrics advertised by the node.

The GCAC actually uses the additive metrics to support the GCAC algorithm chosen for the PNNI protocol. Individual nodes (physical or logical) will need to determine and then advertise the values of these parameters for themselves, based on their internal structure and loading. Note, however, that the PNNI phase 1 GCAC algorithm is primarily designed for CBR and VBR connections; variants of the GCAC are used depending upon the type of QoS guarantees requested and the types of link metrics available, yielding greater or lesser degrees of accuracy. The only form of GCAC done for UBR connections is to determine whether a node can support such connections. For ABR connections, a check is made to determine whether the link or node is authorized to carry any additional ABR connections and to ensure that the ACR for the ABR traffic class for the node is greater than the minimum cell rate specified by the connection.

Using the GCAC, a node presented with a connection request (which passes its own CAC) processes the request as follows [5]:

1. All links that cannot provide the requested ACR, and those whose CLR exceeds that of the requested connection, are "pruned" from the set of all possible paths using the GCAC.

2. From this reduced set, along with the advertised reachability information, a shortest path computation is performed to determine a set of one or more possible paths to the destination.

3. These possible paths are further pruned by using the additive link metrics, such as delay, and possibly other constraints. One of the acceptable paths would then be chosen. If multiple paths are found, the node may optionally perform tasks such as load balancing.

4. Once such a path is found, the node constructs a *designated transit list* (DTL) that describes the complete route to the destination and inserts this into the signaling request. The request is then forwarded along this path. Note that this is only an *acceptable* path to the destination, not the *best* path; the protocol does not attempt to be optimal.

8.2.2.3 Crankback

In a real network, the source-routed path that is determined by a node can only be a best guess. This is because in any practical network, any node can have only an imperfect approximation to the true network-state due to the necessary latencies and periodicity in PTSP flooding. As discussed in Section 8.2.1, the need for hierarchical summarization of reachability information also means that link parameters must be aggregated. Aggregation is a *lossy* process and necessarily leads to inaccuracies.

Each node in the path still performs its own CAC on the routed request because its own state may have changed since it last advertised its state within the PTSP used for the GCAC at the source node. Its own CAC algorithm is also likely to be somewhat more accurate than the GCAC. Hence, notwithstanding the GCAC, there is always the possibility that a connection request may fail CAC at some intermediate node. This becomes even more likely in large networks with many levels of hierarchy, since QoS information cannot be accurately aggregated in such cases. To allow for such cases, without excessive connection failures and retries, the PNNI protocol also supports the notion of crankback.

Crankback is where a connection that is blocked along a selected path is rolled back to an intermediate node, earlier in the path. This intermediate node attempts to discover another path to the final destination, using the same procedure as the original node but uses newer, or more accurate, network state. This is another mechanism that can be much more easily supported in a source-routed protocol than in a hop-by-hop protocol. Figure 8.10 is an example of crankback.

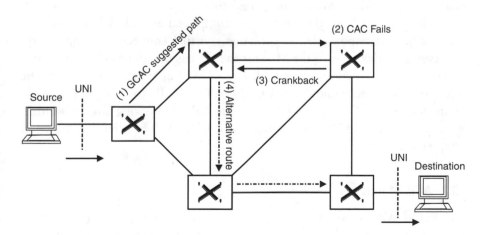

Figure 8.10 Operation of crankback.

8.3 PNNI MIB

In order to configure and control the operation of PNNI protocol, an MIB was defined in [1] and revised in [4]. This MIB is one of the most complicated SNMP MIBs that has ever been developed. This section focuses on the structure of this MIB and the relationship between MIB groups and PNNI protocol. Please refer to the MIB document for detailed definition of each object.

8.3.1 Application of Relevant SNMP Standards

The PNNI MIB conforms to the latest version of SNMPv2 framework—i.e., RFC1902-1908. Specifically, it is defined using the following standards:

- RFC 1902: SNMPv2 SMI;
- RFC 1903: SNMPv2 TC;
- RFC 1904: SNMPv2 Conformance Definitions.

In addition to PNNI MIB, the management of an ATM switching system usually involves other SNMP MIBs. One such MIB is the MIB-II, particularly the system group. The IF MIB (RFC 1573), or at least its interfaces group, should also be supported to provide general information about physical interfaces of the switching system being managed. Each physical interface is modeled as one entry in the MIB-II interfaces table.

The AToM MIB (RFC 1695) is encompassed as well. As mentioned in Chapter 6, it is responsible for permanent virtual connection management. However, a vendor may use the MIB for the management of SVCs if additional management information is provided. PNNI MIB directly refers to the traffic descriptor parameter table defined in the AToM MIB to describe the traffic parameters of SVCC-based RCCs. Also, the AToM Working Group is finishing its MIBs for SVPC/SVCC management, which will be briefly addressed in Chapter 14.

8.3.2 MIB Structure

The PNNI MIB consists of over 20 tables and scalar objects representing dynamic and configured information of the PNNI protocol running over the managed switching system. The structure of the PNNI MIB is depicted in Figure 8.11. A brief description of the functions of each group is given in Table 8.1.

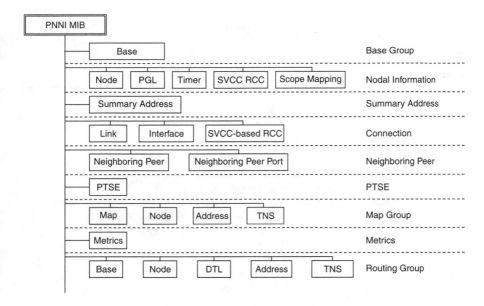

Figure 8.11 PNNI MIB structure.

8.3.3 MIB Groups

The MIB groups defined in the PNNI MIB will be detailed in this section.

8.3.3.1 Base Group

The base group contains overall information about this switching system, including configured parameters and performance statistics. A switching system is defined as a set of one or more physical devices that acts as a single PNNI network management entity. The management information defined can be classified into the following categories:

- PNNI versions negotiation: The lowest and highest versions of the protocol supported at this switching system are defined to support the protocol version negotiation with other systems that support different versions of the PNNI protocol.

- Counters: Two sets of identical counters for routing functions are defined for the switching system acting as an originator and a border node, respectively. In PNNI, every node can generate DTL as an originator. However, only border nodes process crank back connection requests. The specified performance counters collect the total number

Table 8.1
PNNI MIB Group Description

Group	Description	Subgroups
Base	Provides general configuration and performance statistics of the switching system	Base Group
Nodal information	Defines attributes for configuring PNNI node, PG election, timers for protocol operations, SVCC-based RCC, and address scope mapping	Nodal table PGL election table PNNI node timer table Nodal SVCC-based RCC variables table Scope mapping table
Summary address	Contains a list of the summary address prefixes that may be advertised by the specified logical PNNI entity	Summary address table
Connection	Provides additional information necessary to configure and analyze connections, including physical interfaces logical links and SVCC-based RCCs	PNNI interfaces table Link table SVCC-based RCC table
Neighboring peer	Describes the relationship between logical nodes of this switching system and their peer nodes and ports of peer logical nodes	Neighboring peer table Neighboring peer port table
Topological map	Comprises information that can be used to create a topological map of the entire network	Map table Node map table Address map table TNS map table
PTSE	Maintains the node's topology database, which consists of a collection of all PTSEs received	PTSE group
Metrics	Metrics for each connection per direction	Metrics table

Group	Description	Subgroups
Routing	Comprises reachability information, such as routing tables of this system, including precalculated routes to reachable addresses and reachable transit networks in the PNNI routing domain	Base group Node table DTL table Reachable address table Reachable TNS table

of DTL stacks, crankbacks, alternative routes, failures in computing DTL stack, and unreachable destinations.

8.3.3.2 Nodal Information Group

This group contains per-logical node attributes about the switching system being managed. An entry in the nodal table is indexed by the PNNI node index. Other tables are conceptual extensions of the node table, thus augmenting the PNNI node index. Usually, a lowest level switching system only has one entry in these tables, while a PGL has one entry at each level of the hierarchy where it acts on behalf of a PG. However, under one exception, a switching system may support multiple entries in the tables when it is accidentally split into multiple partitions due to failure of equipment. The structure of this group is depicted in Figure 8.12.

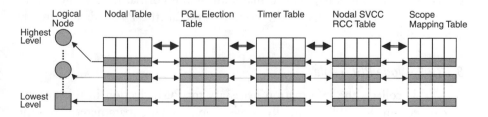

Figure 8.12 Structure for the nodal information group.

8.3.3.2.1 Nodal Table

This table collects nodal configuration and status information that affects the operation of a PNNI logical node. There is a single row in this table for each PNNI PG that the managed system is expected or eligible to join. The managed objects can be classified into three categories as described in the following sections.

Configuration. The configuration information defined in the node table contains nodal addressing information, such as the level of PNNI hierarchy at which this node exists, logical node identifier, routing domain name, ATM address, and PG ID. In addition, there are a number of nodal information flags specified as follows:

- Nodal representation: Indicates whether the node is represented as a complex node or whether it is a lowest level node;
- Restricted branching: Indicates whether the node is able to support additional point-to-multipoint connections;
- Restricted transit: Indicates whether the node is restricted to not allowing support of SVCs transiting this node.

Statistics. Node PTSE's object is defined to gauge the total number of PTSEs currently in this node's topology database.

Status. The operational status and administrative status are used to indicate the desired and actual status of this logical node as discussed in Chapter 2. In addition, the nontransit for PGL election flag is specified to indicate whether the node is operating in a topology database overload state. A node is said to be in a topology database overload state when it is unable to store the complete topology database it is required to maintain by the normal PNNI algorithms, due to insufficient memory in the node to store all the PTSE data currently active in the peer.

8.3.3.2.2 Node PGL Table

This table collects the information necessary for PGL election in this system. The objects in this table essentially come from two parts: parameters from the PGL election data structure and the actual PGL parameters.

PGL election data structure. PNNI specifies a PGL election algorithm for PGL election based on a set of parameters PGL referred to as election data structure defined in [1]. Node PGL group incorporates all these parameters except the search timer specified in the node timer table. The parameters include the node ID for the preferred PGL, preferred PGL leadership priority, the state for the PGL election state machine, and values for various timers used in the algorithm.

Current PGL parameters. This set of parameters includes current PGL node ID, active parent node ID, and a time stamp recording when the current PGL established itself. Besides the configured parent node, an index is specified to identify the node that will represent this PG at the next higher level of hierarchy, if this node becomes PGL.

8.3.3.2.3 Node Timer Table

This table maintains per-node initial PNNI timer values and significant change thresholds in this switching system. The objects are defined according to the architectural variables in the PNNI specification. With this table, a network manager is able to configure all these values to control the frequency of flooding and other interactions of PNNI protocol so as to optimize performance.

Timers. Timer values are defined to control the operation of hello protocol, PTSE exchange, etc. For example, hello protocol uses two timers, the interval timer to control the interval of sending hellos in the absence of triggered hellos and the hold-down timer to limit the rate at which a node sends both triggered hellos. In addition, the hello inactivity factor is used to determine when a neighbor has gone down.

Thresholds. PNNI protocol works in such a way that any significant changes in topology-state parameters may trigger the flooding of PTSEs. This group specifies a set of threshold parameters that are useful in deciding whether a change is significant. The objects include AvcrMt, AvcrPm, CdvPm, and CtdPm. The AvcrMt is the minimum threshold used in the algorithms that determine significant change for ACR parameters, expressed as a percentage. The remaining three parameters are the proportional multipliers expressed as a percentage used to determine significant change for ACR, cell delay variation, and cell transfer delay parameters, respectively.

8.3.3.2.4 Node SVCC-Based RCC Variables Table

This table contains timers and traffic information for SVCC-based RCCs in a switching system.

Timers.   This table keeps a collection of timer values from the PNNI architectural variables [1] for SVCC-based RCCs. The Init/Retry Time is the amount of time this node will delay before initiating establishment or attempting to reestablish an SVCC-based RCC. The calling/called integrity time is the amount of time this node will wait for an SVCC—for which it has initiated establishment as the calling party or has decided to accept as the called party—to become fully established before giving up and tearing it down.

Traffic descriptor.   This is an index into the atmTrafficDescrParamTable defined in RFC 1695. This traffic descriptor is used when establishing switched VCs for use as SVCC-based RCCs to/from PNNI logical group nodes, which implies that the traffic contract for PNNI SVCC-based RCC is symmetric in transmit and receive directions.

It should be noted that the traffic model used in PNNI is consistent with the traffic management 4.0, whereas the model in RFC 1695 conforms to UNI 3.1. Unfortunately, PNNI does not specify exactly how to map traffic parameters of an SVCC-based RCC to the atmTrafficDescrParamTable. However, it will probably no longer be a problem when the new release of AToM MIB is published, since it is likely that the new MIB will conform to traffic management 4.0.

8.3.3.2.5 Scope Mapping Table

A scope is defined as the level of advertisement in the PNNI routing hierarchy for an address, which is specified by a level indicator. PNNI address scoping provides a mechanism for a network administrator to limit advertised reachability information within configurable bounds. This allows network administrators flexible control of reachability information. For example, a reachable address with scope 80 will not be advertised by a PNNI node with scope 72.

The UNI scope for a group of addresses is the highest level PG containing the group address for anycast purposes. Calling parties will not be allowed to establish connections outside of their highest PG. For instance, a user with address scope 80 cannot ring a node with scope less than 80.

The scope mapping table is used to map the UNI 4.0 fifteen level organizational scope values used at UNI interfaces to PNNI routing level indicators. A network-specific mapping is used to translate between the organizational scope indicated across the UNI and PNNI routing level indicators. The mapping applies to both membership scope, which is passed across the UNI via ILMI address registration and is then advertised with the corresponding reachable address in PTSEs as a PNNI routing level, and to connection scope, which is passed across the UNI in a SETUP message. The mapping is configurable via the objects defined in this group. The default values for this mapping are shown in Table 8.2.

8.3.3.3 Summary Address Group

This group has only one table, the summary address table, which lists all the summary address prefixes that may be advertised by the specified logical PNNI entity. A summary address is an abbreviation of a set of addresses, represented by an address prefix that all of the summarized addresses have in common. A logical node can have an arbitrary number of summary addresses.

An entry in this table is identified by both an ATM address prefix and an indicator to indicate the length of the prefix. There are two types of ATM summary addresses: An internal address is the node's native PNNI summary address, while an external address is foreign to that node. Three objects are defined to represent address prefix, address length, and address type, respectively. Sometimes, a node may not want to advertise some of its addresses. The advertisement of a summary address can be suppressed by configuring the *suppress* object. Lastly, a *state* object is used to indicate whether the summary address is currently being advertised by the node within the local switching system into its PG.

Table 8.2
Default UNI Scope to PNNI Level Mappings

UNI Scope	PNNI Routing Level Indicator
1 3	96
4 5	80
6 7	72
8 10	64
11 12	48
13 14	32
15 (Global)	0

Each entry in the summary address table contains only one summary address. The table is indexed by the node index, summary address type, summary address prefix, and the summary address prefix length. A complete list of all summary addresses that belong to a specific node can be obtained by traversing the rows of the table for which the pnniNodeIndex is the same.

8.3.3.4 Connection Group

This group contains connection information of physical interfaces, logical links, ands SVCC-based RCCs . The PNNI interface table provides the information necessary to configure interfaces for PNNI protocol. The link table contains information for logical links, which is automatically discovered by the PNNI protocol. The SVCC table, on the other hand, includes the attributes necessary to analyze the operation of the PNNI protocol on SVCC-based RCCs. The structure of this group is shown in Figure 8.13.

8.3.3.4.1 Interfaces Table

This table contains the information necessary to configure a physical interface of VPCs that have been configured for PNNI's use on a switching system that is capable of being used for PNNI routing. Interfaces may represent physical

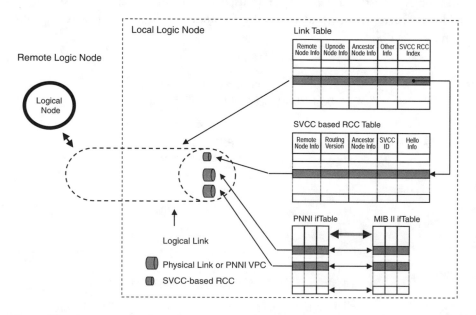

Figure 8.13 Structure of the connection group.

connection points or VPCs that have been configured for PNNI's use. Each interface is attached to a specific lowest level node within the switching system.

One entry in this table is created by the managed system for each row in the ifTable that has an ifType of atm (37) or atmLogical (80). This table is indexed by the ifIndex, which has the same value as the ifIndex object defined in RFC 1573 for the same interface, since this table is a conceptual extension of the ifTable in the IF MIB, containing PNNI-specific interface information. The information includes administrative preferences for the use of each service category over the interface, the characteristics of the RCC, and aggregation attributes.

8.3.3.4.2 Link table

This table lists all the PNNI links attributes necessary to describe the operation of logical links attached to this switching system and the relationship with the neighbor nodes on the other end of the links. A PNNI link is an abstract representation of the connectivity between two PNNI nodes, including individual physical link, individual VPC, parallel physical links, and/or VPCs.

The entire table is read-only as this information is discovered dynamically by the PNNI protocol rather than configured. Each entry represents a link attached to a PNNI logical node in this switching system. As a logical link is identified by both the PNNI node index and the port ID, so is an entry in this table. The management information is defined in this table as follows:

- Link characteristics: Link characteristics include link type, routing protocol version, and the index to the SVCC-base RCC table, which contains full details of the SVCC.

- Connection information: Connection information about the other end of the link, such as remote node ID, remote port ID, and the up node ID, up node ATM address, and common PG ID, has been specified if it is an outside link and up-link.

- Hello: Objects are specified to represent the state of the hello protocol, and the transmitted/received hello packets over this link.

8.3.3.4.3 SVCC-Based RCC Table

The SVCC-based RCC table contains the attributes necessary to analyze the operation of the PNNI protocol on SVCC-based RCCs. The entire table is read-only, as SVCC-based RCCs are established dynamically during operation of the PNNI protocol rather than configured. Each entry in the table includes information about an SVCC-based RCC from a PNNI logical node in this switching system. The table is indexed by the local logical node index and SVCC-based RCC index. Management information is organized into the following categories.

- Connection information: This includes the ID and ATM address of the remote node at which the SVCC-based RCC terminates, the index of the local interface from which the SVCC leaves the switching system, and the VPI and VCI values of the SVCC, which are collectively referred to as SVCC ID.

- Hello protocol information: This consists of the state of the hello protocol, the number of transmitted and received hello packets over this SVCC, and the version of the PNNI routing protocol used to exchange information with the neighbor node.

8.3.3.5 Neighboring PG

This group defines information about neighboring peer entities of a switching system as seen from the perspective of a local node, which consists of the two following tables.

- Neighboring peer table: Contains all the attributes necessary to describe the relationship a node in this switching system has with a neighboring node within the same PG;

- The neighboring peer table: Lists all the ports connected to the neighboring node.

Again the entire group is read-only, as the neighboring peers are discovered dynamically by the PNNI protocol.

8.3.3.5.1 Neighboring Peer Table

This table contains all the attributes necessary to describe the relationship a node in this switching system has with a neighboring node within the same PG. The management information is defined as follows:

- Peer information: This includes the node ID of a neighboring peer and the state of the peer node's state machine.

- Connection information: This contains a count of the total number of ports that connect to the neighboring peer and the index that identifies the SVCC being used to communicate with the neighboring peer in the SVCC-based RCC table.

- Packets exchanged: The neighboring peer table keeps track of all the packets exchanged between the local node and a peer node by defining a number of counters to count the number of following packets:
  - Transmitted/received database summary packets;
  - Transmitted/ received PTSPs;
  - Transmitted/received PTSE request packets;
  - Transmitted/received PTSE acknowledge packets.

A concatenation of the node identifier of the node within the local switching system and the neighboring peer's node identifier is used to form the instance ID for this table.

8.3.3.5.2 *Neighboring Peer Port Table*

This table contains a list of IDs from active ports that is only used when lowest level nodes are connected by physical links and/or VPCs. Each entry represents a port, which is indexed by the local node index, remote node ID, and peer port ID. Two objects are specified in this table. In addition to a port ID, a status object is also defined to indicate whether the port is being used for transmission of flooding and database synchronization information to the neighboring peer.

8.3.3.6 PTSE Group

This group is actually a node's topology database consisting of a collection of all PTSEs received, which represent the node's present view of the PNNI routing domain. As mentioned earlier, PTSEs are the units for distributing information throughout a PG. An entry in the table includes information of only one PTSE in the topology database of a PNNI logical node in this switching system.

The PNNI MIB has specified six types of information contained in PTSEs. They are nodal state parameters, nodal information, internal reachable addresses, exterior reachable addresses, horizontal links, and up-links, which are identified by a *type* object defined in the MIB. This group also specifies objects to represent the sequence number, checksum, and lifetime of the instance of a PTSE as it appears in the local topology database. The PSTE itself is also included in this group.

A concatenation of the node ID of the local node that received the PTSE, the originating node's node ID, and the PTSE ID assigned to the PTSE by its originator are used to form the instance ID for an instance of this object.

8.3.3.7 Topological Map Group

This group comprises the information received from the PNNI protocol that can be used to create a topological map of the entire network. This group consists of four tables: The map table describes connectivity both inside and outside a PNNI node; the node map table provides attributes about logical nodes in the map; the address map table contains reachable address information; and the transit networks table incorporates reachable transit networks per-logic node. Since all information is discovered dynamically during operation of the PNNI protocol rather than configured—by gathering PTSPs—all objects within this group are read-only.

8.3.3.7.1 Map Table

This table contains attributes necessary to find and analyze the operation of all links and nodes within the PNNI hierarchy, as seen from the perspective of a local node.

The map table actually describes the connectivity information for three types of PNNI entities as follows.

- Horizontal link;
- Up-link;
- Node.

The descriptions of these three entities are so similar that only one table is used with the introduction of a *type* object to differentiate these three entities. The drawback is that some of the object names are actually misnomers, as differences do exist.

There are two types of nodal connectivity: Outside connectivity includes horizontal and up-links between nodes, and internal connectivity, on the other hand, describes the connections inside a node. A PNNI node may represent a whole network that has its internal structure modeled as a complex node.

Link-specific information.   The description of outside connectivity is straightforward. The originating node ID and port ID identify the originating point of the link as do the remote node ID and port ID for the remote end point of the link. In addition, another object is defined to contain the derived aggregation token value for this link.

Normally, there are two rows in the table for each horizontal link. These two rows provide information for a network management application to map port identifiers from the nodes at both ends to the link between them.

Node-specific information. For a PNNI node, the important information includes the originating node ID and the input and output port pair, as defined in the nodal state parameters information group in [1].

Common information. Objects common to both internal and outside connectivity include the following.

- PG ID of the originating node;
- PTSE ID originated by the originating node that contains the information group(s) describing the PNNI entity;
- Indicator to indicate whether VPCs can be established across the PNNI entity being described;
- Metrics tag as an index into the metrics table that will be described in Section 8.3.5.8;
- Index into the set of link and nodal connectivity associated with the originating node and port in order to distinguish possible multiple connections.

This table is indexed by the local node index, originating node ID, originating port ID, and the index. Figure 8.14 illustrates an originating node and a remote node connected by two links. Each link is identified by a different index. However, the original and remote node/port fields for both connections are the same.

The map table provides necessary information to describe the map of the network. Figure 8.15 is an example of traversing a connection using this table.

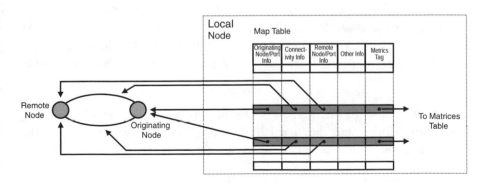

Figure 8.14 Connectivity using map table.

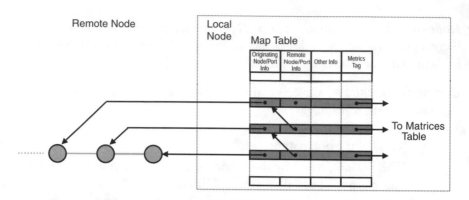

Figure 8.15 Topological traversing using map table.

A network manager starts from one entry in the table, in which the original node fields contain the information of the originating node. The remote node fields, on the other hand, contain the pointer to the next node linked by this connection. The connectivity information of the next node can be obtained by following the pointer. By repeating this procedure, the topological information of the whole network from the perspective of a local logical node can be found.

8.3.3.7.2 Nodal Map Table

This table lists all nodes as seen from the perspective of a local node. An entry in the table contains *vertical* information about a node in the PNNI routing domain. The local node index and the remote node ID index this table. This table is actually defined according to the nodal information group specified in [1].

Originating node information.   This includes the node ID, PG ID, ATM address, and the four flags described in Section 8.3.5.2.1—the nodal representation flag, restricted branching flag, restricted transit flag, and PGL election flag.

Originating node PGL information.   The objects are the preferred PGL, the leadership priority, and the "I am leader" flag, which indicates whether the originating node claims to be PGL of its PG.

Originating node's parent information.   This includes the parent node ID, parent ATM address, parent PGL node ID, and Parent PGL node ID.

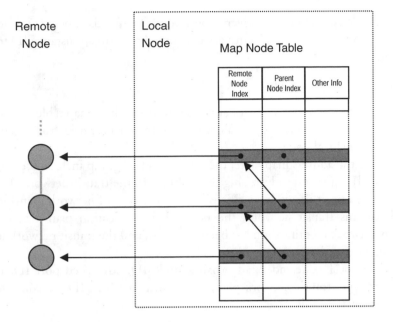

Figure 8.16 Node map table.

The nodal map table is organized in such a way that the hierarchy of the remote nodes can be easily followed. In addition to the attributes of the remote node, a pointer to its next higher level (i.e., its parent node) is also included in the table. This makes it possible for the network manager to traverse the whole hierarchy from the originating node to the highest level, which is depicted in Figure 8.16.

8.3.3.7.3 Address Map Table

This table contains a list of all reachable addresses from each node visible to the local node. An entry in the table contains information about an address prefix reachable from a node in the PNNI routing domain, as seen from the perspective of a PNNI logical node in this switching system.

The objects specified are the reachable ATM prefix (including the address prefix and prefix length), the node ID of the advertising node, and the port identifier used from the advertising node to reach the given address prefix. Since a node can advertise multiple reachable addresses, an additional index is used to enumerate all of the addresses advertised by the specified node.

The local node index, advertising node ID, advertised port ID, reachable address, and address prefix length are combined to form an instance ID for this object.

8.3.3.7.4 TNS Map Table

The structure of this table is similar to that of the address map table. The difference is that it contains a list of all reachable transit networks from each node visible to the local node.

Like the address map table, the objects in this group include the node ID and port ID of a node advertising reachability to the transit network. In addition, three transit network-specific objects—the type of network identification used for this transit network, the network identification plan according to which network identification has been assigned, and the transit network identifier—are defined.

The local node index, advertising node ID, advertised port ID, transit network type, transit network plan, and transit network ID are combined to index an entry in this table.

8.3.3.8 Metrics Group

This group has only one table that contains information of available resources either associated with a PNNI entity or with the connectivity between a PNNI node and a reachable address or transit network. This information is used to attach values of topology state parameters to nodes, links, and reachable addresses.

Each entry contains a set of parameters that applies to the connectivity from a certain node and port to another node or port or to one or more reachable address prefixes and/or transit networks, for one or more particular service category(s). This is indicated by an object that acts as a bit-mask with each bit that is set representing a single-service category for which the resources indicated are available. The topological state parameters consist of nine objects—the administrative weight and metrics 1 to 8—as described in Section 8.2.2.2. A GCAC indicator is also defined to indicate whether the advertised GCAC parameters apply for CLP=0 traffic or for CLP=0+1 traffic.

This table is indexed by a tag that is used to associate a set of topological parameters that are always advertised together, the direction in which the parameters in this entry apply, and an index into the set of parameters associated with the given tag and direction.

Figure 8.17 illustrates the usage of administrative weight. This is the main metric used by PNNI for computation of paths. The assignment of

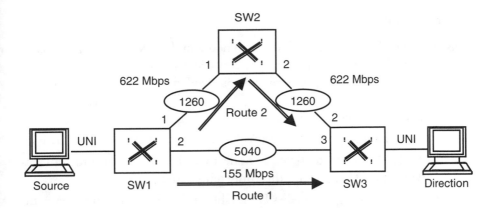

Figure 8.17 Example of administrative weight.

administrative weights to links and nodes influences how PNNI selects paths in the private ATM network. Since it is an additive topological metric, the resulting administrative weight for a path is simply the sum of the individual weights of the links along the path. An ATM switch selects paths with the least administrative weight when they satisfy the requested QoS of a connection. The administrative weight indicates the relative preference of a link assigned by the private network owner. For instance, it might depend on link capacity or link length. Administrative weight can also be used to exclude certain links from routing, such as a backup link that needs to be used only when the primary link is full. The administrative weight for the links in Figure 8.17 are configured according to the link speed. The higher the speed, the lower the administrative weight. Thus, route 2 is selected over route 1, which is a direct connection, because route 2's administrative weight (2520) is less than that of route 1 (5040).

8.3.3.9 PNNI Routing Group

This group comprises reachability information that consists of precalculated and configured routes to reachable addresses and transit networks in the PNNI routing domain.

8.3.3.9.1 Routing Base Group

This base group provides general information about the whole routing group. It has two objects: one to count the total number of current precalculated PNNI routes to valid PNNI nodes and the other to count current PNNI routes from nodes in the PNNI routing domain to addresses and transit networks.

8.3.3.9.2 Routing Node Tables

This table contains precalculated routes to nodes in the PNNI routing domain. Each entry includes a particular route to a particular destination node under a particular policy. An entry in this table is jointly identified by the following.

- Local node index;
- Service category with which this forwarding table entry is associated;
- Node ID of the destination node to which this route proceeds, at which the DTL stack for this route terminates;
- Entry index into the node's DTL table of the DTL stack that goes with this route.

Topological state information for both forward and backward directions has also been defined in the same format as in the metrics table. The parameters are described in Section 8.2.2.2.

Moreover, there are other objects defined to provide additional information including the following:

- Port ID of the destination node at which the route terminates;
- Routing mechanism by which this route was learned (e.g., from PNNI, management, etc);
- The time at which this route was last updated or otherwise determined to be correct;
- A reference to MIB definitions specific to the particular routing protocol that is responsible for this route;
- An indicator that notes whether any advertised GCAC parameters apply for CLP=0 traffic or for CLP=0+1 traffic.

Figure 8.18 illustrates how various fields are used to represent the routes to a destination node. A detailed description of a route is given in the DTL table, which will be introduced in Section 8.3.3.9.3.

8.3.3.9.3 DTL Table

This table contains all DTL stacks used for the precomputed routes maintained by this managed entity. Each entry includes a segment of a DTL stack. A

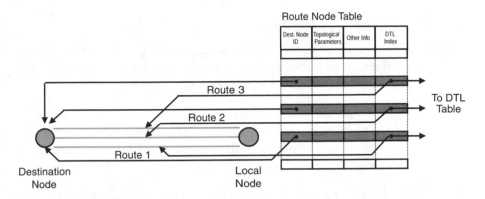

Figure 8.18 Route node table.

complete DTL stack is formed by traversing the rows of the table for which the DTL index is the same.

A stack segment consists of the node of this hop in the DTL stack, the port from which this hop exits, and the type of link out from this node that can be used to indicate level transitions in the PNNI hierarchy.

The table is indexed by the PNNI owner's node index, the DTL index that identifies the DTL stack in this node's DTL table, and the entry index that describes an entry in the current DTL stack. The simplified operation of getting DTL from the DTL table is depicted in Figure 8.19. Note that the node ID and port ID can be used jointly to determine the next node in the DTL list.

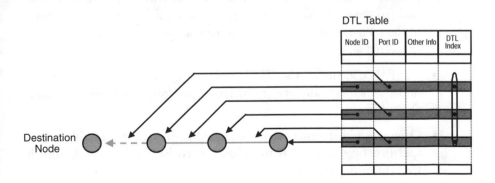

Figure 8.19 DTL table.

8.3.3.9.4 Reachable Address Table

This table includes all addresses that this switching system believes reachable. It is also used to configure static routes to reachable address prefixes. Each entry in the table contains information about a reachable address prefix.

Advertising node information.   This includes the local interface over which the reachable address can be reached, the node ID of a node advertising reachability to the address prefix, and the port identifier used from the advertising node to reach the given address prefix.

Address attributes.   Eight objects are defined to describe the attributes of the reachable address prefix:

- Type indicates whether the reachability from the advertising node to the address prefix is internal or exterior.

- Protocol indicates whether the reachable address advertised is automatically discovered by the PNNI protocol or configured by a network management station.

- PNNI scope of advertisement indicates the level of reachability from the advertising node to the address prefix.

- VPC capability indicates whether VPCs can be established from the advertising node to the reachable address prefix.

- Metrics tag is an index into the metrics table for the topological state parameter values that apply for the connectivity from the advertising node to the reachable address prefix. There will be one or more entries in the metrics table whose first instance identifier matches the value of this variable.

- PTSE ID identifies the PTSE being originated by the originating node, which contains the information group(s) describing the reachable address.

- Originate advertisement indicates whether or not the reachable address specified by this entry is to be advertised by the local node into its PNNI routing domain.

- Address information makes reference to MIB definitions specific to the particular routing protocol that is responsible for this reachable address prefix.

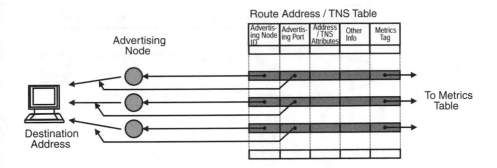

Figure 8.20 Route address/TNS table.

Status. In addition to the row status, there are two status objects. The operation status object indicates whether the reachable address prefix is operationally valid and whether it is being advertised by this node. The time stamp object, on the other hand, indicates when the connectivity from the advertising node to the reachable address prefix became known to the local node.

This table is indexed by the local node index that received the reachability information, reachable address, address prefix length, and an index that distinguishes between multiple listings of connectivity to a given address prefix from a given local node. The operation of this table is illustrated in Figure 8.20.

8.3.3.9.5 Reachable Transit Network Table

This table includes all transit networks that this switching system believes reachable. It is also used to configure static routes to reachable transit networks. The structure of this table is exactly the same as that of the reachable address table. The only changes are that all address prefix-specific objects are replaced with a set of transit network-specific objects.

8.3.3.9.6 Example of Using Route Group

The roles of each table in this group in PNNI routing are illustrated in Figure 8.21. The address table and TNS table provide necessary routing information from the last hop in the PNNI hierarchy, that is, from a destination system to the nodes that advertise reachability to the desired destination. The node table contains routing information from the local node to the advertising nodes, while the DTL table contains the precomputed DTL stacks to route a call to the advertising nodes.

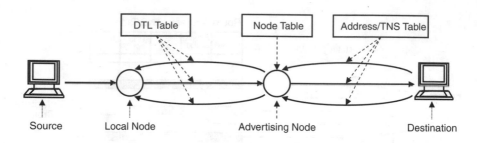

Figure 8.21 Operation of PNNI routing group.

The following is a set of simplified steps to find a source-routed route for setting up a call:

1. Look up the reachable address table or the transit network table by using the destination address prefix to find out the nodes that advertised reachability to the desired destination, as well as topological state information and other attributes that are associated with these nodes.

2. Look up the routing node table by using the advertising node IDs and service categories requested by the user to find out the routes to the advertising nodes, as well as topological state information and other attributes that are associated with these routes.

3. Compare all the routes to the desired destination using the GCAC algorithm to select a suitable one in terms of the topological information retrieved from these tables, as already introduced in Section 8.2.2.3.

References

[1] ATM Forum, "Private Network Node Interface Specification Version 1.0," (af-pnni-0055.000), March 1996.

[2] ATM Forum, "PNNI v1.0 Addendum (Soft PVC MIB)," (af-pnni-0066.000), September 1996.

[3] ATM Forum, "PNNI v1.0 Addendum (ABR parameter negotiation)," (af-pnni-0075.000), January 1997.

[4] ATM Forum, "PNNI v1.0 Errata and PICS," (af-pnni-0081.000), May 1997.

[5] Anthony Alles, "ATM Internetworking," Cisco Systems, Inc., http://www.cisco.com/warp/public/614/12.html.

[6] Castineyra, I., N. Chiappa, and M. Steenstrup, "The Nimrod Routing Architecture," RFC 1992, August 1996.

9

LAN Emulation

9.1 Introduction

A key to the success of ATM will be the ability to allow for interoperability between existing networking technologies and ATM, given the vast installed base of LANs today and the network and link layer protocols operating on these networks. To facilitate the evolution of the network from current LANs to ATM, it is essential that ATM networks support the existing applications and the upper level protocol stacks currently in use.

However, the ATM architecture differs fundamentally from legacy LAN technologies [1, 2]. ATM is a connection-oriented technology, in which a VC is set up before exchanging any data between two ATM stations. Legacy LANs, on the other hand, employ connectionless transmission technology. Another significant difference between LANs and ATM networks is that legacy LANs and their associated protocols were developed to fully exploit the characteristics of the shared media used—particularly the broadcast/multicast transmission—which are intrinsically supported in LANs. In contrast, the connection to each host from another host is logically separate from all other connections in ATM networks, even if a continual piece of cable is used between all the hosts on a LAN. While this approach has many benefits, such as increased security and better possibilities for sharing of bandwidth, one drawback is that ATM no longer supports broadcast or multicast. Moreover, the addressing models are different. The addressing space for LANs is flat, whereas ATM is hierarchical.

LANE over ATM defined by the ATM Forum provides the means to resolve the differences between these two technologies. As the name suggests, LANE software typically resides in a *driver* below a standard operating system-defined MAC-level interface within an ATM-connected computer to emulate the behavior of the driver.

From the viewpoint of network administration, constructing an *emulated LAN* (ELAN) out of LANE has significant advantages over legacy LANs. ELANs greatly simplify network operation and maintenance by letting network administrators group users based on common interests rather than physical location. Network administrators can add, move, and change an ELAN simply by redefining groups in the NMS and remotely configuring software in the end device or ATM switch. No change of cabling or adding of equipment is required.

9.2 LANE UNI

9.2.1 Architecture

The LANE specifications [1–5] concentrate on defining the protocol used across the *LANE UNI* (LUNI) between LECs and the LANE service.

The main objective of the LANE, as stated in [1, 2], is to "enable existing applications to access an ATM network via protocol stacks like APPN, Net-BIOS, IPX, AppleTalk etc., as if they were running over traditional LANs." LANE works at the *media access control* (MAC) layer and enables legacy Ethernet, token ring, or FDDI traffic to run over ATM with no modifications to applications, network operating systems, or desktop adapters. Legacy end stations can use LANE to connect to other legacy systems, as well as to ATM-attached servers, routers, hubs, and other networking devices. This group of devices is logically analogous to a group of LAN stations attached to an Ethernet or token ring segment. An ELAN looks like a normal LAN in every respect except for bandwidth as far as either end systems attached to the LAN ports on the LAN switches or the higher layer protocols operating within the ATM hosts or routers are concerned. Their operation does not differ in any manner.

An ELAN supported by the LANE is a common user group that is only able to exchange similar types of data frames within the same broadcast domain. Many ELANs may exist concurrently in the same ATM network, but they cannot communicate directly; hence, a router or bridge is required. The LANE protocol stack is shown in Figure 9.1. With LANE, existing applications running over ATM still use the same interfaces as those used in legacy LANs, specifically the *network driver interface specification* (NDIS) and the *open data-link interface* (ODI).

ELANs operate in SVCs, PVCs, or a combination of both. The current specification only details operations for 802-type frames (802.3 and 802.5). The aim of the LUNI is to hide the complexities of call set-up and to provide emulated services similar to the "bus-" like broadcast operations of conventional LAN technologies.

LUNI specifies components and protocols to emulate the characteristics of a real LAN, namely the broadcast/multicast services and the connectionless data transmission services. In addition, it defines the *LANE address resolution protocol* (LE_ARP) to convert the flat MAC addresses to the hierarchical ATM ones.

9.2.1.1 Components

LANE is based on the well-known client/server model. Each ELAN is composed of a group of LECSs and a LANE service as depicted in Figure 9.2. A LANE service consists of three different functional entities: the LECS, *LANE server* (LES), and the *broadcast and unknown server* (BUS). These three services may exist independently but are typically co-located in a single device. For instance, LE services may reside in ATM switches, routers, bridges, or workstations. The functions for each entity are as follows:

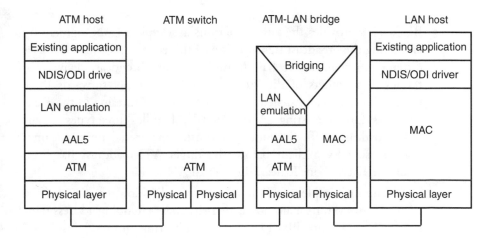

Figure 9.1 LANE protocol stack.

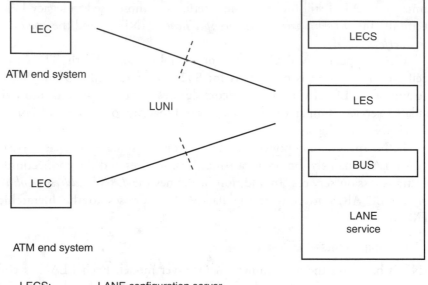

LECS:	LANE configuration server
LEC:	LANE client
LES:	LANE server
BUS:	Broadcast and unknown server
LUNI:	LANE UNI

Figure 9.2 LUNI.

- BUS: BUS provides the same functions as a typical LAN segment by emulating the broadcast transmission of this segment; however, it uses different mechanisms to handle multicast and broadcast delivery of frames to members of the group. The BUS simply forwards these frames directly to all its clients and then relies on filtering in those clients as is done in existing LANs. The BUS handles three types of traffic—broadcast traffic, all multicast traffic, and initial unicast frames that are sent by a client before the target ATM address has been resolved. There is one logical BUS per ELAN.

- LES: The LES implements the control coordination function for the ELAN. However, its major job is to assist clients in doing address resolution. The LES provides a facility for registering and resolving MAC addresses and/or route descriptors to ATM addresses. There is also one logical LES per ELAN.

- LEC: The LEC is the entity embedded in an ATM attached end system that performs data forwarding, address resolution, and other control functions within a single ELAN. LEC is identified by a unique ATM address and is associated with one or more MAC addresses reachable through that ATM address. Although the current LANE specification defines two types of ELANs, one for Ethernet (802.3) and one for token ring (802.5), it does not permit direct connectivity between an LEC that implements an Ethernet ELAN and one that implements a token ring ELAN. Two such LECSs can only be interconnected through an ATM router that acts as a client on each ELAN. LECSs can also be divided into two types in terms of their roles. A proxy LEC is one that represents the MAC addresses of devices other than itself. In other words, it acts as a bridge. A non-proxy LEC is a device like a host that has a unique MAC address of its own.

- LECS: The LECS is responsible for the automatic assignment of individual LECS to ELANs. It assigns any LEC that requests configuration information to the most appropriate ELAN service by giving the LES's ATM address to the client based upon its own policies and configuration database, as well as information provided by LECS. A single LECS can manage the configuration information for a very large ATM network, as its responsibilities are limited to initial configuration.

9.2.1.2 Connections

In order to exchange control and data traffic between the components mentioned above, LANE defines separate connections for different purposes. These connections may be PVCs, SVCs, or mixed SVCs/PVCs.

9.2.1.2.1 *LECS-LEC*

Configuration direct VCC. This bi-directional VCC is set up between an LEC and its LECS by the LEC as part of the LECS connect phase and is used to obtain configuration information, including the ATM address of the LES. The LEC may continue to keep it open for further queries to the LE Configuration Service while participating in the ELAN.

9.2.1.2.2 *LES - LEC*

Control direct VCC. This bi-directional point-to-point VCC is set up between the LES and an LEC by the LEC as part of the initialization phase.

Since the LES has the option to use the return path to send control data to the LEC, this requires the LEC to accept control traffic from this VCC. The LEC and LES must maintain this VCC while participating in the ELAN.

Control distribute VCC. This point-to-point or point-to-multipoint control VCC between LES and an LEC by the LES may be optionally set up to distribute control traffic as part of the initialization phase. If set up, the LEC is required to accept the control distribute VCC regardless of type. The LES and LEC must maintain this VCC while participating in the ELAN.

9.2.1.2.3 BUS-LEC

Multicast send VCC. This bi-directional point-to-point VCC between LEC and BUS is set up by the LEC that first issues LE_ARP requests and receives the LE_ARP response. This VCC is used for sending multicast data to the BUS and for sending initial unicast data. The BUS may use the return path on this VCC to send data to the LEC, so this requires the LEC to accept from this VCC.

Multicast forward VCC. This point-to-multipoint or uni-directional point-to-point from the BUS to its LECS is set up by and used by the BUS for distributing data from the BUS. The BUS may forward frames to an LEC on either the multicast send VCC or the multicast forward VCC. Thus, the LEC must be able to accept frames on either VCC.

9.2.1.2.4 LEC-LEC

Data direct VCC. This is a bi-directional point-to-point VCC between LECSs that want to exchange unicast data traffic. The role of this VCC is to emulate the connectionless services by making use of the connection-oriented ATM link.

Figure 9.3 is a summary of all LANE components and connections.

9.2.2 Protocol Operation

This section explains how different LUNI components work together to implement LANE, from the time an LEC first joins an ATM network until it exchanges data with its peers. This process typically begins when an LEC initiates communications with the LECS and registers its interest in joining an ELAN. The operation of the LANE is divided into several phases in terms of the interactions between LEC and the LAN service as described below.

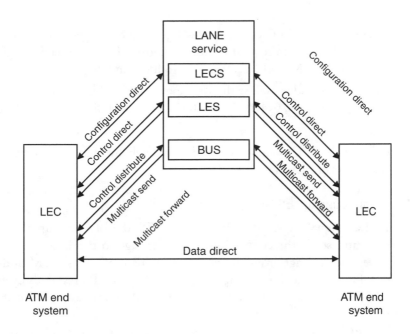

Figure 9.3 LANE connections.

9.2.2.1 Configuration

When an LEC first powers up, it must obtain configuration information from the LECS in order to join an ELAN. The LEC sets up a configuration direct VCC and sends the "LE_CONFIGURE REQUEST" frame with some useful "Initial State" information, such as its ATM address, MAC address, preferred LAN type, and maximum frame size. The LECS examines information provided in the request, assigns the LEC to the most appropriate ELAN based upon its own policies and configuration database, then responds with a "LE_CONFIGURE RESPONSE," containing the actual LAN type, the actual maximum frame size, and the ATM address of an LES. By providing an LES address, the LECS implicitly assigns the LEC to an ELAN.

The LANE specification [2] specifies three mechanisms for an LEC to find the ATM address for an LECS. They are listed as follows in the order in which an LEC should try them:

- Preconfigured: The LECS address for an LEC may be configured by the network manager. The LEC should try to establish a configuration direct VCC first, if it is configured with an LECS address.

- Via ILMI: The LEC can send ILMI get-request or get-next-request to its ATM switch to request the LECS address from the service registry table of the ILMI MIB as already introduced in Chapter 5.
- Well-known Address: Use the well-known LECS address defined by the ATM Forum, i.e., 0x470079000000000000000000-00A03E000001.

Note that the LECS can be bypassed completely by manually configuring the ATM address of LES in the LEC.

9.2.2.2 Connecting to LES/BUS

Once an LEC knows the ATM address of the LES, the LEC sets up a control direct VCC and sends the connection set-up message to the LES. When the LES receives the message from the client, it learns the LEC's ATM address from the calling party field in the message. Typically, the LES responds by adding the LEC as a leaf node on the point-to-multipoint control distribute connection.

The LEC then registers its MAC address and associated ATM address with the LES; in return, the LES assigns the client an LEC ID in the LE_JOIN_RESPONSE after successfully validating the parameters in the request. The ATM and MAC addresses are stored by the LES for future reference in address resolution. At this point, the LEC can resolve MAC addresses to ATM addresses. The first address it needs is that of the BUS.

The LEC sends an LE_ARP_REQUEST message to the LES requesting the ATM address associated with the all-ones MAC address (i.e., the broadcast address). The format of this request looks like any other LE_ARP message. The LES responds with the ATM address of the BUS in the LE_ARP_RESPONSE.

Once the client obtains the BUS address, it sets up a multicast send connection to the BUS; the BUS, on the other hand, observes the ATM address of the client. Like the LES, the BUS typically adds the LEC as a leaf node in the point-to-multipoint Multicast Forward Connection. The LEC is now ready for data transfer.

9.2.2.3 Data Transfer

In this phase, an LEC may transmit two types of frames as detailed below.

Unicast Frame

When an LEC receives a unicast data frame for transmission, it first checks its local LE_ARP cache entries to see whether it knows the ATM address associated with the MAC address. If there is no such entry, it cannot immediately set up a data-direct VCC to the target. The LEC should initiate an ARP process by sending an LE_ARP_REQUEST to its LES over the control direct VCC. The LE_ARP message includes the source ATM address of the LEC making the request. The LES may respond with the information obtained when that LAN destination was registered or simply forward the LE_ARP to all clients via the control-distributed VCC.

The target client recognizes the MAC address and sends an LE_ARP_RESPONSE to the LES, which includes both its own ATM address and the source ATM address for the LEC originating the LE_ARP request. The LES broadcasts the response message with the target ATM address to all the LECS. The cycle ends when the originating LEC recognizes its own ATM address contained in the response. At that point, it has learned the ATM address associated with the unknown MAC address and can set up a data-direct connection to the target LEC. If the source LEC receives subsequent frames with the newly learned MAC address, it immediately forwards them down the data-direct VCC.

Each LEC builds up its own LE_ARP cache entry table to cache the MAC addresses, ATM addresses, and VCC bindings. If a particular MAC address has not been active for some time, an LEC will eventually drop it from its cache. When there are no more MAC addresses associated with a data-direct VCC, the LEC can drop the connection.

Though the LEC can hold the data frame until it learns the ATM address of the target from LE_ARP_RESPONSE, it may choose to forward the frame to the BUS while initiating the response. The BUS then forwards it to every client in the ELAN. This is the only way such a frame can be assured of reaching its destination if it is behind a bridge that has not yet learned this particular address [1].

Broadcast or Multicast Frame

If an LEC receives a broadcast or multicast frame, it immediately passes the frame to the BUS via the VCC established initially. The broadcast or multicast frame can be identified by its destination MAC address. A leading 1 indicates a broadcast or multicast message, while zero indicates a unicast message.

The BUS, in turn, forwards that message to all nodes on its point-to-multipoint BUS forward VCC; that is, all LECSs on the ELAN receive the broadcast. If the BUS receives two broadcast or multicast frames simultaneously, it buffers one briefly while it sends the other one out. This serialization prevents intermixing of cells from different data frames on the VCC to the

LECS. AAL5 allows an LEC to reassemble only one data frame at a time on a single VCC.

Protocol information, particularly the LEC ID assigned by the LES, in the header of every frame uniquely identifies the LEC that originated the broadcast to the BUS. To speed up broadcast frame processing, an LEC checks the LEC ID for every incoming frame; if there is a mismatch, the LEC disposes of the frame immediately. Otherwise it looks at the destination address. A non-proxy LEC saves only those frames whose destination MAC address matches its own multicast address. Proxy LECS save all multicast frames.

9.3   LANE Management Framework

LANE emulates the characteristics of a legacy LAN segment. In essence, however, LANE and LAN are totally different technologies. As such, the management of LANE is totally different from that of legacy LANs. Unlike the traffic in legacy LAN segment that is transmitted over a shared medium, the ATM traffic is hard to monitor. LANE traffic is spread out over many virtual circuits instead of being concentrated in one physical network segment. To make things worse, these virtual circuits may be set up and torn down frequently, presenting a moving target to would-be observers. Moreover, the performance of each virtual circuit may be affected by factors outside the control of the ATM ELAN hosts. In particular, these include which switches are congested and what actions those switches take in response to congestion (cell loss, flow control, etc.). In addition to data traffic, there may be a fair amount of LANE control traffic from a server that is both a potential bottleneck and a single point of failure. It is desirable to monitor this control traffic separately. Furthermore, it is an explicit non-goal for LANE to support promiscuous listeners (hosts who want to listen to all unicast traffic). Thus, managers cannot simply attach protocol analyzers to an ELAN to find out what is happening.

LANE defines standards to allow dynamic and sophisticated management of ELANs by distributing management functions among LANE components and other relevant MIBs. The ATM Forum has specified four MIB modules for LANE management.

The client management [4] involves only one MIB module, the LEC MIB, which specifies LEC management information.

Service management [5] encompasses three MIB modules:

- ELAN MIB: Provides ELAN configuration and LECS management information;

- LES MIB: Covers LES management information;
- BUS MIB: Provides BUS management information.

Service management defines three MIB modules since there is no guarantee that any of the components in question will be co-located. Usually an LES and BUS are likely to be managed by a single agent. However, it is highly possible that the user of information in the ELAN MIB (e.g. an LECS) will not be co-located with the LES or BUS. Moreover, the LECS may not be implemented at all in some ELANs. Thus, three MIB modules are needed for the management of three servers separately.

The procedures to create an ELAN that a client can join will actually require three stages using these MIBs:

1. Create a new ELAN in the ELAN MIB;
2. Create an LES for that ELAN using the LES MIB;
3. Create a BUS for that ELAN using the BUS MIB.

The configuration of LANE components to an ELAN (i.e., LECS, LES, BUS and LEC) is done in the MIB groups listed in Table 9.1.

It is important to understand that the ELAN MIB only contains a repository of static information—for example, which clients should be assigned to which LESs and BUSs. In order to determine the current topology of an ELAN—i.e., which clients are attached to which LESs and BUSs—the LES MIB and BUS MIB (and the LEC MIB if detailed information on the client is required) must be used.

Table 9.1
MIB Groups for the Configuration of LANE Components

LANE Component	MIB	Group
LECS	ELAN MIB	LECS Configuration table
LES	ELAN MIB	LES table
BUS	LES MIB*	BUS table
LEC	ELAN MIB	ELAN Policy table
		LEC assignment tables

* This is due to the fact that LECs learn the BUS address only from the LES to which they correspond.

Although the management functions vary from one LANE component to another, they share the same management framework. The LANE management framework adopts the party-based SNMPv2 as the basis for defining LANE MIBs. However, it does not mandate the use of SNMPv2, as opposed to SNMPv1.

Besides the MIB modules defined in the LANE specifications, other MIB modules should also be used, particularly the MIB-II interfaces group or its evolved form—RFC-1573 IF-MIB—and the ATM MIBs. LECS and servers use VCs (switched or permanent) to communicate with one another. LE MIBs provide indexes for these VCCs, so that a network manager can obtain the attributes of the VCCs from the ATM MIBs. Specifically, the AToM MIB is useful in managing PVCs; while the supplemental MIB [6] from the AToM MIB working group is capable of SVCC management. In addition, proprietary MIB extensions may by used.

All SNMP agents supporting LESs may implement MIB II or IF MIB and are encouraged to implement RFC 1695, whereas it is mandatory for SNMP agents that support LEC to implement these two MIBs.

9.4 Client Management

The LEC management is specified in the ATM Forum's standard, "LAN Emulation—Client Management Specification, Version 1.0" [4].

9.4.1 Management Model

9.4.1.1 Management Tasks

The major objectives of the LEC management are as follows [4]:

- Examine initial state parameters, including the LAN name and LES ATM address for each operational client. Given that an LES is likely to know all its current clients, identifying the LES may make it easier for a network manager to locate other clients.
- Monitor statistics for LANE control traffic and SVC failures. This can indicate whether an LEC is being swamped by LE_ARPs or is failing to receive them
- Examine LE_ARP caches.

The LEC MIB also provides the optional abilities to perform the following functions.

- Create and destroy LECSs;
- Configure a client's initial state parameters;
- Create and destroy LE_ARP cache entries.

Note that the management of the data direct VCCs, the identification of LANE PVCCs, or identification of the ATM addresses at each end of a VCC is expected to be handled via extensions to other MIBs.

9.4.1.2 Emulated Network Interfaces and the Interfaces Table

Each real LAN port has an entry in the MIB-II Interfaces table and the RFC 1573 interface extensions table respectively. To be consistent with the existing management framework, the LEC management specification [4] requires that each emulated port has entries in these tables as well.

To identify an ELAN, it must be tagged with one of these ifType constants:

- aflane8023(59): For an ELAN that supports the IEEE 802.3/Ethernet data frame formats;
- aflane8025(60): For an ELAN that supports the IEEE 802.5/token ring data frame format.

These values let a network management application know that additional information about the interface is available via the LEC MIB.

It should be noted that the entry for an LEC in the MIB-II interfaces table or the RFC 1573 interface extensions table models two interfaces as illustrated in Figure 9.4: the emulated 802.3/802.5 packet interface between the LEC and higher (sub)layers and the VCC-oriented interface between the LEC and the AAL5 (sub)layer. The reason why there is one interfaces table row per LEC is that the LEC-to-AAL5 interface does not have enough interesting MIB-II-style traffic measurements to justify a separate interfaces table entry.

The mapping between an ATM port and an LEC is not required to be one-to-one. That is, a single ATM port may support several active LECSs, and a single LEC may employ several ATM ports. In order to describe the multiplexing relationship, the stack table in the IF-MIB may be used.

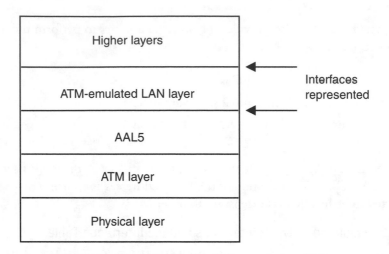

Figure 9.4 LANE interfaces modeled in the interfaces table.

Unlike most of the other interfaces, the LEC interfaces can be created and destroyed by network management depending on the status of LECS. Most tables in the LEC MIB use a separate LEC index (lecIndex) so that the agents can control the ifIndex allocation. As a result, a special MIB group, the index mapping group, is defined to translate the ifIndex to the LEC index and vice versa. Note that the LEC index is used to identify an instance of an LEC throughout this MIB.

9.4.2 MIB Structure

The LEC MIB is organized into five groups in terms of their respective functionalities, as depicted in Figure 9.5.

Note that the connection management in LEC MIB is limited to the connections between an LEC and its servers (i.e., LECS, LES, and BUS). The LEC MIB does not provide any mechanisms to identify data direct VCCs, to get VCC topology information (local-end and far-end ATM addresses) that is not currently in the AToM MIB, or to configure PVCs for LANE use. It is expected that some of these features will be incorporated into other ATM MIBs (see Chapter 14 for more information).

9.4.3 Index Mapping Group

As mentioned earlier, LECSs can be created and destroyed dynamically by network management. Just like an ordinary Ethernet or token ring interface, an

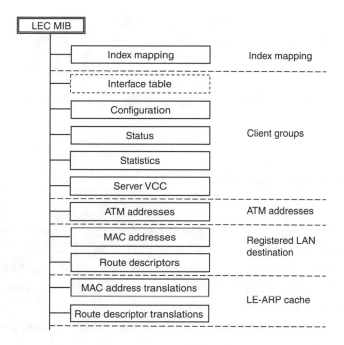

Figure 9.5 LEC MIB structure.

LEC has a row in the MIB II/IF-MIB interfaces table, which is indexed by its ifIndex. Traditionally, ifIndex values, which are chosen by agents, are permitted to change across restarts. However, using ifIndex to index a LANE interface in an LEC could complicate row creation and/or cause interoperability problems if each agent had special restrictions on ifIndex. Having a separate index, i.e., LEC Index, to identify avoids these problems.

The index mapping group describes the mapping between ifIndex and the lecIndex. It consists of only one table, the LEC mapping table, with one row for each LEC. This table is indexed by the ifIndex. The only columnar object is the LEC mapping index, which identifies the corresponding LEC index value. Figure 9.6 illustrates the mapping between the ifIndex and the lecIndex using the index mapping table. The mapping of lecIndex to ifIndex is provided in the LEC status table, which will be described in Section 9.4.4.2.

9.4.4 LEC Group

This group encompasses five tables, which are specified to provide overall information for each LEC in this system as follows:

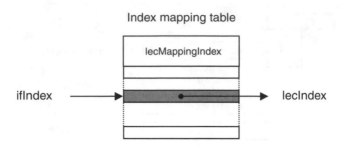

Figure 9.6   ifIndex to lecIndex mapping.

1. MIB-II interfaces table: Describes the emulated packet interface an LEC presents. This group is specified in the MIB-II or IF MIB. The detailed mapping of objects onto an LEC interface is given in [4].

2. LEC configuration table: Contains configuration variables used to create, delete, and configure LECSs.

3. LEC status table: Contains client status and operational parameters.

4. LEC statistics table: Contains various performance statistics.

5. LEC server connection table: Contains information about control and multicast VCCs.

All the tables in this group are indexed or augmented by the lecIndex rather than by ifIndex as explained earlier. The structure of this group is illustrated in Figure 9.7.

9.4.4.1 LEC Configuration Table

The LEC configuration table is responsible for the creation, deletion, and configuration of LECSs in a host, with one row for each LEC. Unlike hardware ports, network managers can create/delete an LEC interface dynamically. However, the RFC 1573 interfaces table does not directly support row creation. Therefore, creating or deleting a row in the LEC configuration table is defined to have the side effect of creating or deleting corresponding rows in the following tables.

- MIB-II/RFC 1573 interfaces table;
- LEC index mapping table;
- LEC status table;

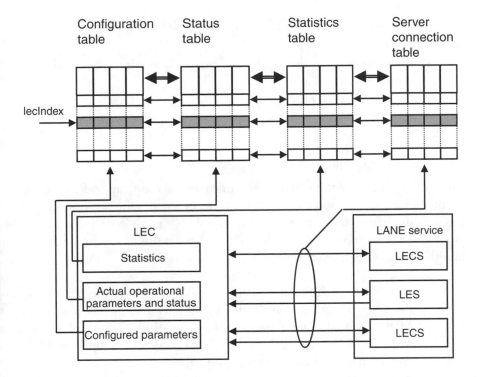

Figure 9.7 LEC client group structure.

- LEC server VCC table;
- LEC statistics table;
- Any other dependent tables.

Most objects in this table correspond to the settable initial state parameters, which are annotated with the appropriate (Cxx) as specified in the LANE specification [1, 2, 4]. For example, C5 is the LAN type, which is either 802.3 or 802.5. The information includes the configured key parameters for configuration/join phase (the actual operational parameters are listed in the LEC status table) as well as other parameters that may be changed either by the network manager or by the LECS.

In addition, there are other objects defined for LEC identification, creation, and destruction as follows:

- LEC index: Identifies the client;

- Row status: Used to create and destroy clients on hosts that support this;
- Owner: The entity that configured this entry and is therefore using the resources assigned to it;
- Configure mode: Controls whether this client uses the LECS to auto-configure.

9.4.4.2 LEC Status Table

This group contains current status, identification, and operational parameters for the LECS that this host supports. It consists of a table with one row for each LEC. Again, many objects correspond to initial state parameters and are annotated with the appropriate (Cxx) label. The objects in this group can be divided into three categories listed as follows:

1. Identification: Contains the primary ATM address and LEC ID of the LEC, as well as the ifIndex to identify the client's row in the MIB-II/RFC 1573 interfaces table. The ifIndex object allows the mapping of lecIndex back to the ifIndex, which is the opposite of the operation of the LEC index mapping table.

2. Status: Includes operational state of the client, last failure information, LANE protocol supported by the client, and protocol version. These objects provide LEC-specific status information regarding an LEC, which is not included in the MIB-II ifTable.

3. Operational parameters: Current parameters used by the LEC, including the ATM addresses of its LES and LECS, ELAN type, ELAN name, maximum data frame size, and an indicator as to whether the client is a proxy client. Note that the maximum data frame size of an LEC is different from the ifMtu in the ifTable. The value of the latter is a function of the former and the LAN type, which is defined in [4].

9.4.4.3 LEC Statistics Table

This group consists of a table that contains various statistics collected by an LEC from the LANE control traffic and SVCs, with one row for each LEC. The management information can be classified into the following categories:

- LE_ARP: Counters of the LE_ARP requests and replies received and transmitted by this client;

- Control frames: The total number of control frames sent and received by this client;
- SVC failures: The number of SVCs that this client either tried but failed to establish or rejected for protocol or security reasons.

9.4.4.4 LEC Server Connections Table

This table provides per-LEC indexes into ATM MIB(s) for control and multicast VCCs. Each VCC is identified by three parameters: ifIndex, VPI, and VCI, as discussed in Chapter 6. The VCCs managed by this group are listed as follows:

- LEC-LECS: Configuration direct VCC;
- LEC-LES: Control direct VCC and control distribute VCC;
- LEC-BUS: Multicast send VCC and multicast forward VCC.

9.4.5 LEC ATM Addresses Group

Each LEC has zero or more ATM addresses, one or more addresses if it is operational. These addresses may denote different ATM ports or the same port. Several LECS may share an ATM port, provided that they use different ATM addresses. The addresses in the LEC MIB are organized as follows:

- Primary ATM address: LEC status table;
- Additional ATM address(es): ATM address table.

This group only consists of the ATM address table, containing an ATM address of the client and a row status object to allow the network manager to create and delete table rows. The LEC index and ATM address jointly index the table.

9.4.6 Registered LAN Destination Group

LANE currently supports both Ethernet and token ring emulation. In Ethernet emulation, a LANE component need examine only a data frame's destination MAC address in order to direct the frame towards its ultimate destination(s). In token ring emulation, however, a LANE component may have to use a "route descriptor" extracted from the data frame's *routing information field* (RIF), instead of the destination MAC address, to properly direct the frame over the ELAN. For the sake of simplicity, LAN destination is used instead of the destination MAC address or a route descriptor. Throughout LANE MIBs, LAN destination for Ethernet is represented by only one object, i.e., MAC address; while the route descriptor needs two objects: segment ID and bridge number.

The registered MAC addresses table and the registered route descriptors table list all of the local unicast MAC addresses (initial state parameter C6) registered for this host's LECS that support Ethernet and token ring, respectively. Both tables have an identical structure in which objects denoting a LAN destination and the ATM address registered for the LAN destination are defined. Both the LAN destination and LEC index are used to identify an entry in the tables. LAN destination for Ethernet is represented by only one object (i.e., MAC address), while the route descriptor needs two objects, segment ID and bridge number. Each row describes a binding between a LAN destination and ATM address pair registered for a particular LEC, as depicted in Figure 9.8.

9.4.7 LE_ARP Cache Group

LE_ARP cache is a table of entries, each establishing a relationship between a unicast or multicast LAN destination external to the LEC and the ATM address to which data frames for that LAN destination will be sent. The LE_ARP MAC address table holds the mapping between MAC addresses and their corresponding ATM addresses. The route descriptor LE_ARP cache table, on the other hand, perseveres the mapping between route descriptors and ATM addresses. Both the LEC index and LAN destination ID jointly identify an entry in the tables.

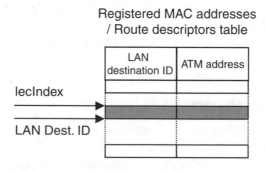

Figure 9.8 Structure of the registered LAN destination tables.

The common objects defined in both tables include a LAN destination and its corresponding ATM address, the type of the entry, and a row status used to create and delete of an entry in the table. There are five different ways in which an entry in these two tables is created, reflecting by the entry type object:

1. learnedViaControl: Learned by using the LE_ARP_REQUEST/ LE_ARP_RESPONSE protocol or by otherwise observing relevant traffic on control VCCs. Entries of this type are aged.

2. learnedViaData: Learned by observing incoming traffic on data VCCs. Entries of this type are aged.

3. staticVolatile: Created by the network manager, which is static and will be lost after a restart of the agent;

4. staticNonVolatile: Also created by network manager, it is static and will survive a restart of the agent;

5. other: Does not fall into one of the categories defined above.

In addition to the management information described above, another object is defined to indicate whether a MAC address is local or remote in the LE_ARP MAC addresses table only. Figure 9.9 shows the structure of LE_ARP MAC addresses table and LE_ARP route descriptor table.

9.5 Service Management

LANE service management is defined in the "ATM Forum, LANE—Server Management Specification, Version 1.0" [5].

9.5.1 Management Functions

The objectives of the LANE service management MIBs are to allow the network manager to configure and to monitor the operation of various components. The specification defines a list of detailed goals in terms of the configuration management, performance management, and fault management as follows:

- Configuration management:
  - Create and destroy ELANs;

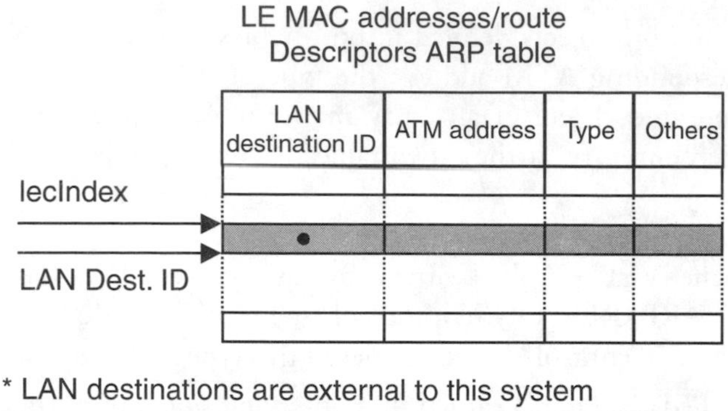
LE MAC addresses/route
Descriptors ARP table
LAN
destination ID
ATM address
Type
Others
lecIndex
LAN Dest. ID
* LAN destinations are external to this system

Figure 9.9 Structure of the LE destination ARP tables.

- Assign clients to and delete them from an ELAN;
- Control the parameters of the LECS, LES, and BUS;
- Monitor ELAN topology (which LECS are joined to an LES);
- Identify the VCCs being used by the LECS, LES, or BUS.

- Performance management:
    - Statistics for each ELAN server;
    - Statistics for each ELAN server-LEC pair.
- Fault management:
    - Operational status for each service component;
    - Error logs for each service component.

Unlike LECSs, LESs present no interfaces as defined in MIB-II. They are, however, related to one or more ATM interfaces in many ways:

- LECSs can be listening to one or more ATM interfaces with a well-known ATM address as defined in the LANE specification.
- LES can be receiving/sending control or ATM LE-ARP traffic via one or more ATM interfaces.
- BUS can be receiving/forwarding traffic on one or more ATM interfaces.

The LANE service management MIBs are designed to facilitate the accommodation of future LANE server-to-server protocol management extensions, which will allow multiple servers—multiple LESs, for example—to coexist in a single ELAN. However, no objects have been specified to manage the interactions between these servers.

9.5.2 LANE Server Management Model

Although the roles of the LANE servers LECS, LES, and BUS vary, management of the different servers share many similarities. The MIB groups defined to manage each server are shown together in Figure 9.10. It is apparent from Figure 9.10 that there is a significant portion that is common to all three servers. Server-specific MIB groups, on the other hand, reflect the differences between servers.

Specifically, the VCC tables and fault management groups for the three servers are exactly the same. The configuration tables and the statistics tables are similar to each other.

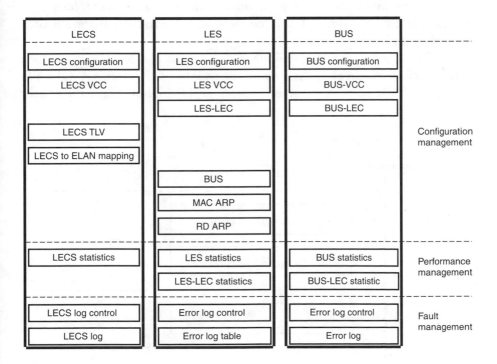

Figure 9.10 LES management model.

Furthermore, the identification of the servers uses the same mechanism. An instance of a server is identified by a server index such as LEC index, LES index or LECS index. In order to ensure the uniqueness of the index, a read-only scalar object, namely the server configuration next ID, is used to provide the next available server index by the agent.

9.5.3 ELAN MIB

9.5.3.1 Structure

An ELAN is defined by the ELAN name, a set of TLVs, and other parameters in terms of network management. The structure of this MIB is depicted in Figure 9.11, which consists of two portions, ELAN and LECS.

The assignment of an LEC to an appropriate ELAN is the major role of this MIB, which is implemented by rules known as *policies.* The ELAN MIB simply deals with information that is required to enable a client to join an ELAN. This information can be divided into two categories, information required in deciding which ELAN a client should join and information needed for a client to join the ELAN.

This information would typically be used by an LECS in deciding which ELAN a client should be assigned to on the basis of its CONFIG_REQUEST. It is possible for an ELAN to exist without an LECS, however, with a client

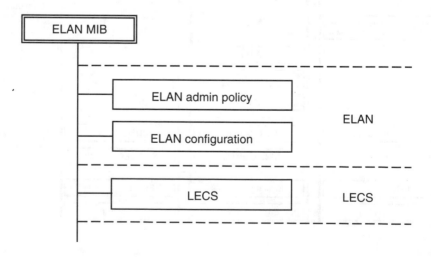

Figure 9.11 Structure of the ELAN MIB.

obtaining this information by some other means. LANE is designed to allow the ELAN MIB to exist without requiring an LECS. This is why the LECS portion of the ELAN MIB is a separate group. Systems that implement the LECS must implement the LECS group, to allow the LECS to be monitored and controlled. TLV information also resides in this part of the MIB since this is specific to the LECS configuration mechanism.

9.5.3.2 ELAN Administration Group

This group actually only provides a registry for the LEC assignment policy types. The six most commonly used LEC assignment policies are defined in this group:

1. By ATM address: Assign an LEC to an ELAN by its ATM address;

2. By MAC address: Assign an LEC to an ELAN by its MAC address;

3. By route descriptor: Assign an LEC to an ELAN by its route descriptor;

4. By LAN type: Assign an LEC to an ELAN by its LAN type, i.e., ieee802.3 or ieee802.5;

5. By packet size: Assign an LEC to an ELAN by its data frame size;

6. By ELAN name: Assign an LEC to an ELAN by its ELAN name.

The definitions above represent the set of assignment policy types that may be supported by the standard implementations of MIBs. The assignments are based on the information available in a typical CONFIG REQUEST. Vendors may specify their own policies by defining similar tables or objects in their own MIBs.

9.5.3.3 ELAN Configuration Group

This group provides configuration information for ELANs. The structure of the group is depicted in Figure 9.12, consisting of eight tables.

An ELAN is constructed or destroyed in this group for configuration purposes only (i.e., an ELAN can be created and many LECS can be assigned to this ELAN by policies). However, this MIB would not reflect whether LECSs have actually joined the ELAN. The network manager has to poll the individual LECS or the LES/BUS to determine the actual ELAN topology.

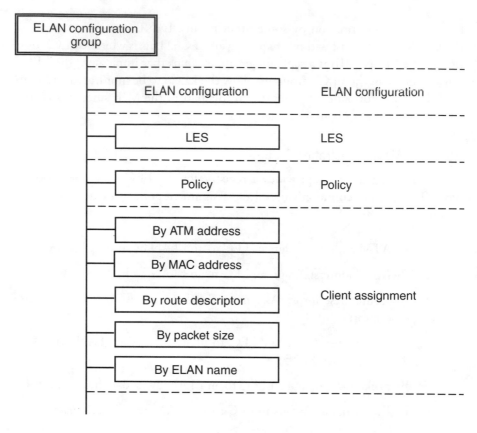

Figure 9.12 Structure of the ELAN configuration group.

9.5.3.3.1 ELAN Configuration Table

This table lists all ELANs that the agent manages. Each row in this table represents an ELAN. ELANs should be created or destroyed from this table in conjunction with operations to create or destroy the related service components.

This table can be used by the entity that assigns LECS to ELANs. This might be an LECS, or in an environment where there is no LECS, the information might be used directly by a system to configure LECS.

The objects in this table are specified according to the initial state parameters [1, 2], including the configured ELAN name, LAN type, maximum frame size, and a TLV selector to enable it to choose a set of predefined TLV properties. The TLV selector is set to zero if LECS is not supported, or there is no TLV associated with the entry.

This table is indexed by the ELAN configuration index object, which identifies a particular ELAN. In order to ensure the uniqueness of the index, a read-only scalar object, namely the ELAN configuration Next ID, is used to provide the next available ELAN index by the agent.Figure 9.13 depicts the structure of LECS configure table.

9.5.3.3.2 ELAN LES Table

The ELAN LES table contains all LESs for each ELAN specified in the ELAN configuration table. This table is used for configuration of an ELAN only; that is, creating an LES in this table does not instantiate an LES in the network. It is done in the LES MIB. Only the ATM address of the LES is held in an entry. If LECS is supported, the value of this object is the LES ATM address that LECS returns to the LEC in the CONFIGURE response. If LECS is not supported, the value of this object pertains to the LES ATM address that the network manager provides to the LEC.

Each ELAN can have more than one LES providing LANE services. Each LES, however, can service only one ELAN. As a result, an LES is indexed by the ELAN configuration index and the ELAN LES index that uniquely identifies an LES. The structure of the ELAN LES table is depicted in Figure 9.14.

9.5.3.3.3 ELAN Policy Table

The ELAN policy table contains all policies this agent supports for assigning an LEC to an ELAN. A set of policies with the same or different priorities can be selected by the entity that provides ELAN configuration service, such as the LECS.

This table is indexed jointly by both a selector index and a policy index. The policy index uniquely identifies a policy, and the selector index allows

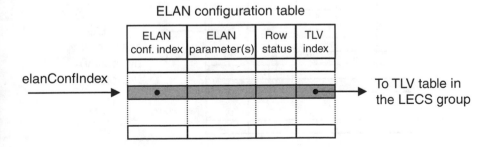

Figure 9.13 LECS configuration table.

ELAN LES table

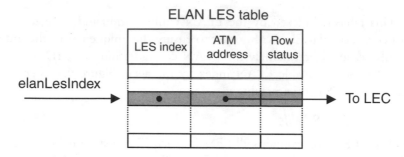

Figure 9.14 ELAN LES table.

multiple policies be selected by one LECS or an entity that is providing ELAN configuration service.

Each policy is associated with its priority and type. When LECS receives a configure request, it checks its policies selected from this table to determine which ELAN and LES the LEC will join. The policy with the highest priority or with the smallest value of the ELAN policy priority is evaluated first. The policies with the same policy priority are evaluated at the same time with the AND operation.

The syntax of the policy type is the autonomous type so that an agent can use both standard policies defined in the ELAN administration group and proprietary policies defined by vendors. Figure 9.15 illustrates the ELAN policy table.

ELAN policy table

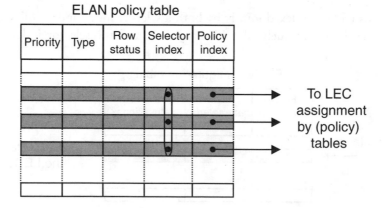

Figure 9.15 ELAN policy table.

9.5.3.3.4 LEC Assignment Tables

An LEC assignment table is used to assign an LEC to an ELAN by a specific policy. The ELAN MIB specifies five such tables:

1. LEC assignment table by ATM address;
2. LEC assignment table by MAC address;
3. LEC assignment table by route descriptor;
4. LEC assignment table by packet size;
5. LEC assignment table by ELAN name.

The LECS or other entity that serves the LANE configuration function looks up these tables to determine the ELAN membership.

All LEC assignment tables share the same structure. Each table consists of a row status object and the objects denoting parameter(s) used by a particular policy, as described in Table 9.2.

Each entry in an LEC assignment table represents an LEC to ELAN/LES binding. Consequently, the table is indexed jointly by the ELAN configuration index that points to the ELAN to which this LEC belongs, the ELAN LES index that points to the LES this LEC should join, and the policy-specific object(s). Figure 9.16 shows the ELAN LEC assignment by policy table.

Table 9.2

Parameters Used in the LEC Assignment Tables

LEC Assignment Table	Policy-Specific Parameter(s)
By ATM address	ATM address
	ATM address mask
By MAC address	MAC address
By route descriptor	LAN ID portion of the IEEE 802.5 route descriptor
	Bridge Number portion of the IEEE 802.5 route descriptor
By packet size	Maximum AAL-5 SDU size this LEC can support
By ELAN name	Name of the ELAN to which this LEC belongs

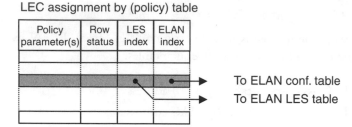

Figure 9.16 ELAN LEC assignment by policy table.

9.5.3.3.5 Client Assignment Procedure

Having introduced the objects used in the ELAN configuration, it is time to
briefly discuss the procedure by which an LEC is assigned to an appropriate
ELAN/LES. The assignment can be carried out by an LECS or whatever entity
that is capable of using the information. Let us begin with an LECS that
receives a CONFIGURE_REQUEST from an LEC.

When an LECS is supported, the LECS assigns a client to an ELAN
based on the information given in the CONFIG request. Each LECS is associated with a set of predefined policies in the ELAN policy table, which is identified by the policy selector index in the LECS configuration table, as will be
discussed shortly in the LECS portion of the ELAN MIB. This allows dis-
jointed policy sets to be created, which makes it possible for individual LECSs
(or other users of this data) to use different policies. The information in the
CONFIG request is checked against the various LEC assignment tables in the
ELAN MIB according to the policy selector index.

If there can be multiple policies used in determining the LEC assignment
at one time, these policies are executed in order of their priorities, with the lowest number getting the highest priority. If multiple policies of the same priority
exist, they are executed at the same time, and all policies must succeed.

Users of the information in the policy and assignment tables should use
the following procedure [5] when attempting to assign clients:

1. Find the first policy with the highest priority level with the appropri-
    ate policy selector index.

2. Check to see if the client matches any assignments for this policy type.

3. If the client passes, there will be one or more ELAN/LES index pairs.

4. If the client fails to find any appropriate ELAN/LES pair with this policy, it has failed and must restart from the next priority down.

5. If there are no policies with a lower priority than the current one, then the client is not assigned to any ELAN and has been rejected.

6. If there are more policies with the same priority level as the one which just passed, these must also be evaluated as described above.

7. If a client has passed all policies with a particular priority level, there will have been at least one ELAN/LES pair generated per priority.

8. If no pair is common to every policy (i.e. the client passed every policy, but the policies failed to agree on any ELAN/LES pair), then the client has failed for this priority level. The behavior is as if the client had failed against any one of the policies at this priority.

9. If a single ELAN/LES pair can be identified as common to the results of all the policies, then this is taken as the final result.

10. If there are multiple matches and the client could be assigned validly to more than one ELAN/LES, any may be chosen, the particular outcome being implementation-defined. Thus, the LECS behavior is implementation-dependent.

Figure 9.17 depicts how various MIB tables are used together to implement the LEC assignment procedure.

A simplified example is given below to illustrate the above procedure. Assume that the LECS has the policy selector index pointing to the policies in the ELAN policy table as shown in Table 9.3, which consists of two priority levels. The highest level contains only one policy entry—by MAC address. There are two policy entries in the second level—by ATM address and by packet size.

When the LECS receives the CONFIG request from an LEC, it follows this procedure:

1. It checks the MAC address against the LEC assignment by MAC address table. Assume there is no match in the table, the first policy fails.

2. Then, the LECS checks the second level, starting from the by ATM address table, as shown in Table 9.4. If the ATM address matches the ATM address prefix for electronic & electrical department, there are four ELAN/LES pair entries in the table. They are ELAN-EE-1516/LES-EE-1516, ELAN-EE-4544/LES-EE-4544, ELAN-EE-9234/LES-EE-9234, and ELAN-EE-18190/LES-EE-18190.

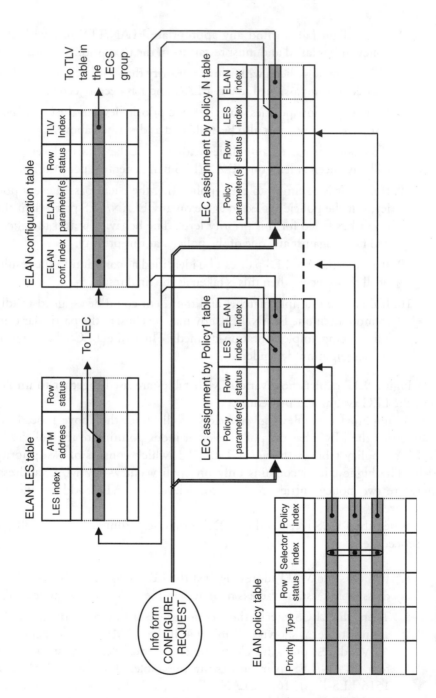

Figure 9.17 LEC assignment.

3. Next, check the second entry in the ELAN policy table. There is one more policy with the same priority level as the by ATM address policy, that is, the by packet size table. If the packet size in the CONFIG request is 4544, there are two matched entries, ELAN-EE-4544/ LES-EE-4544 and ELAN-CS-4544/ LES-CS-4544, as shown in Table 9.4.

4. There is only one ELAN/LES pair that is common to all policies, the ELAN-EE-4544/ LES-EE-4544. Hence, this pair is the choice of the assignment.

Table 9.3
Policy for the Example

Priority	Policy
1	By MAC address
2	By ATM address
	By packet size

Table 9.4
LEC Assignment Tables for the Example

		By ATM Address	
		ATM Address Prefix for Electronic & Electrical Department	ATM Address Prefix for Computer Science Department
By packet size	1516 bytes	ELAN-EE-1516 /LES-EE-1516	ELAN-CS-1516 /LES-CS-1516
	4544 bytes	ELAN-EE-4544 /LES-EE-4544	ELAN-CS-4544 /LES-CS-4544
	9234 bytes	ELAN-EE-9234 /LES-EE-9234	ELAN-CS-9234 /LES-CS-9234
	18190 bytes	ELAN-EE-18190 /LES-EE-18190	ELAN-CS-18190 /LES-CS-18190

9.5.3.4 LECS Groups

The remaining portion of the ELAN MIB is the LECS Groups. It contains management information that enables network managers to configure LECSs and monitor their performance. It should be implemented only when the LECS is implemented. The structure of this portion of the MIB is depicted in Figure 9.18.

9.5.3.4.1 LECS Configuration Group

This group consists of four tables: the LECS configuration table, the LECS to ELAN mapping table, the LECS TLV table, and the LECS VCC table.

LECS configuration table.   This table contains the configuration and status information for all LECSs managed by this agent, which is used by network managers to create, delete, or configure an LECS.

The most important information contained in this table is the LECS connection information that indicates the ATM address and interface to which the LECS is listening for CONFIGURE requests. The network manager can specify an ATM address and its mask as a portion of the address. The actual address

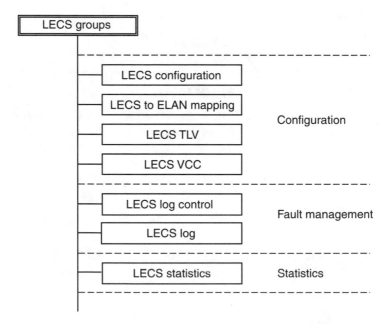

Figure 9.18 Structure of the LECS groups.

is the result of the specified ATM address, its mask, and interaction through the ILMI with the switch, which is represented by the actual address object.

The table also includes an LECS policy selector index pointing to the set of policies used by this LECS in determining the requester's ELAN membership. The policies are defined in the ELAN policy table.

In addition, this table specifies a number of status objects including a time stamp that indicates when this LECS is last initialized and the three statuses, i.e., operational status, administrative status, and row status. The roles of these status objects have already been described. Figure 9.19 illustrates the LECS configuration table.

LECS to ELAN mapping table. This table manages the binding between ELANs and LECS. Each entry represents an ELAN to LECS mapping, which is indexed by both the ELAN configuration index and the LECS configuration index. The only element contained in a row is the RowStatus, which indicates whether there is a mapping between the ELAN/LECS pairing.

LECS TLV table. TLV stands for the type, length, and value or the tag, length, and value. It is a metadata format for generic, self-describing, byte-packed, streamed, aggregate data objects. This table is used to configure TLVs for each ELAN. The table is indexed by a selector index, which allows more than one TLV to be selected by an ELAN; the TLV tag, which specifies the type of the TLV; and a TLV index that is used to distinguish between different entries with the same TLV tag. Network managers can create TLVs and assign them to an ELAN by specifying corresponding selector indexes in both the TLV table and the ELAN configuration table. The LECS TLV table is depicted in Figure 9.20.

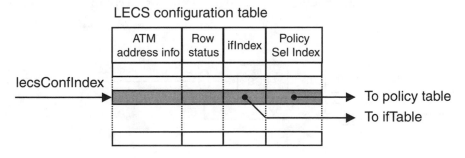

Figure 9.19 LECS configuration table.

LECS TLV Table

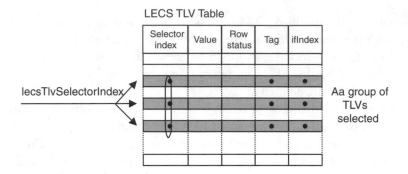

Figure 9.20 LECS TLV table.

LECS VCC table. This table lists the configuration direct VCCs from all LECSs to the LECS, which is designed for the network manager as the central place to trace VCCs. The table is writable if PVC is used and read-only if SVC is used. Each entry represents a configuration direct VCC between a pair of LEC and LECSs. The objects in each entry are defined in such a way that they can uniquely identify a VCC in the AToM MIB—i.e., the ifIndex, VPI, and VCI. These objects along with the LECS configuration index jointly index a row in this table, which is illustrated in Figure 9.21.

9.5.3.4.2 LECS Performance Management Group

This group is used for performance management of all LECSs this agent manages. It contains only one table, the LECS statistics table. This table augments the LECS configuration index, as it is an extension of the LECS configuration table. Each entry in the table maintains counters that record various statistics of the whole LECS. Specifically, each configuration error except the "Version not

LECS VCC Table

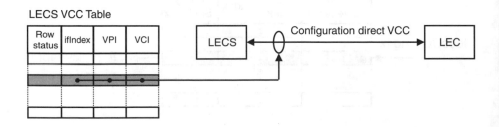

Figure 9.21 LECS VCC table.

Supported" listed in Table 13 of [1] has a corresponding counter in the table. The objects are listed as follows:

- Successfully granted CONFIGURE requests;
- Malformed CONFIGURE requests dropped by the LECS;
- Invalid request parameter error;
- Insufficient resources to grant request error;
- Access denied error;
- Invalid requester-id error;
- Invalid destination error;
- Invalid ATM address error;
- LEC not recognized error;
- LE_CONFIGURE error;
- Insufficient information error.

9.5.3.4.3 LECS Fault Management Group

In addition to operational status objects provided by other groups, LECS fault management is accomplished mainly by keeping error logs of all LECS instances. This group consists of two tables: the LECS fault control table and the LECS error log table. The logged events are saved in the latter, while the former is used to control the error logging. The implementation of this group is optional.

LECS error log control table.   This table contains error log control information of all LECS instances managed by this agent. The information is used to enable or disable error logs for an LECS, as well as to reset or configure the error log of an LECS. Since this table is an extension of the LECS configuration table, it augments the latter. The roles of each object in this table are listed as follows:

- Clear log: To clear the error log entries associated with this LECS;
- Max entries: Maximum entries of the error log an LECS can support;
- Last entry: To point to the last entry in the error log table associated with this LECS;
- Admin status: To enable/disable error logging for the LECS;
- Operational Status: Status of the error log operation.

LECS error log table.   This table contains error logs of the LECS instances enabled in the LECS error control table. Each entry describes when the error occurred, the nature of the error, and the ATM address of the client whose CONFIG request resulted in the error. This table is indexed by the LECS instance index (LECS configuration index) and an error log index.

The error log index for the first entry after reset or clearing the error log is an assigned value ($2^{32}-1$). Succeeding entries are assigned with descending values consecutively. This allows a network manager to retrieve the most recent N entries by using the objects defined in the above two tables. For example, if the number of max entries for the error log is 100 and the last entry is 96, the values of the error log index for the five most recent entries in the error log table are 100, 99, 98, 97, and 96 with the entry 96 being the most recent error. Figure 9.22 illustrates the operation of LECS error log control table and the LECS error log table.

9.5.4 LES MIB

The LES management model is designed to be compliant with the LANE 1.0 specification [1]. This MIB may be used to determine the distribution of the LECs among LESs and to create, configure, and monitor LESs. The MIB is divided into three groups as depicted in Figure 9.23.

- Configuration group: Provides configuration and topology information;

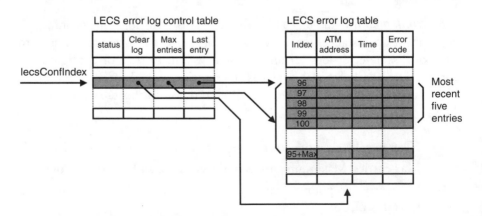

Figure 9.22 LECS fault management.

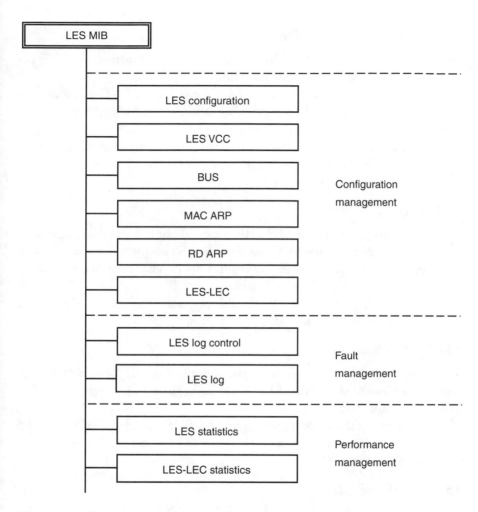

Figure 9.23 LES MIB structure.

- Performance group: Provides various counters for each LES and per-LES/LEC pair;
- Fault management group: Logs LES error information.

9.5.4.1 LES Configuration Group

This group consists of five tables that contain LES configuration information.

9.5.4.1.1 LES Configuration Table

This table lists all LESs managed by this agent. There can be multiple LESs per ELAN, but an LES can serve only one ELAN.

The LES configuration table contains the configured ATM address, the actual address to which the LES is listening to receive ATM ARP requests, and the status objects to control and monitor the LES operation.

The LES configuration table also includes information of ELAN name, LAN type, max frame size, and the time out period used for timing out most request/response control frame interactions.

9.5.4.1.2 LES VCC Table

The VCC management in the LES MIB is carried out in two tables. This table lists all the control distribute VCCs used by the LES to distribute control traffic to the participating LECs, while the control direct VCCs are maintained for each LES-LEC pair in the LES-LEC table.

This structure of the LES VCC table is exactly the same as that of the LECS VCC table, which is introduced in Section 9.5.3.4.1.

9.5.4.1.3 LES BUS Table

As mentioned earlier, the LECs learn the address of the BUS from their LES. This is implemented by the LES BUS table, which contains the ATM address(es) of the BUS(s).

Each entry in this table represents an LES/BUS pair identified by the LES configuration index that identifies an instance of LES and the BUS configuration index that uniquely identifies a BUS ATM address.

9.5.4.1.4 LES ARP tables

The LES ARP table by MAC address and the LES ARP table by route descriptor provide information to map a LAN destination to its ATM address. Both tables have an identical structure. The table contains entries for unicast addresses and the broadcast address. The entry for the broadcast address provides the ATM address of a BUS. When an entry is for a unicast address, the corresponding ATM address is for an LEC.

This table is indexed by the LAN destination of the BUS or an LEC and the LES configuration index that identifies the LES instance.

The information contained in this table includes the corresponding ATM address, the type of the entry, the LEC ID assigned by the LES during the join phase, and a row status used to create and delete of an entry in the table.

The MAC to ATM address mapping can be obtained via three different ways reflected by the entry type object as follows:

- viaRegister: Registered by the LEC;
- staticVolatile: Created by the network manager, which is static and will be lost after a restart of the agent;
- staticNonVolatile: Also created by the network manager, it is static and will survive restart of the agent.

Compared with the LE ARP table by the MAC address table in the LEC MIB, the LES ARP table by MAC address does not include the object that indicates whether the address is remote or not. This information is contained in the LES-LEC topology table, which will be discussed in Section 9.5.4.1.5. There is also an extra object in LES ARP table that the LE ARP table does not have, that is, the LECID.

9.5.4.1.5 LES-LEC Topology Table

This table contains the LES to LECs mapping (i.e. the topology information of an ELAN). The agent creates an entry in this table when an LEC registers successfully with the LES. The management information of the LEC includes the identification of the control direct VCC (ifIndex, VPI, and VCI) used to connect the LES, LECID, ATM address, and an indicator showing whether the LEC is a proxy or a normal one. There are three status objects defined to indicate the operational status, row status, and the last change time stamp for the LEC. Note that the control distribution VCCs are managed by the LES VCC table.

The table is indexed jointly by the LES LEC index, which identifies the LEC, and the LES configuration index, which identifies the instance of the LES to which the client joins. Figure 9.24 illustrates the different roles of the objects defined in the LES-LEC table in describing topological information of an ELAN. The control direct VCCs between LES and LECs are identified by the ifIndex, VPI, and VCI, while the LECs in the ELAN are identified by the LEC IDs.

Figure 9.25 summarizes LES VCC management, which encompasses two tables. The LES VCC table provides information of the point-to-multipoint control distribute VCC, whereas the LES-LEC table contains pair-wised information for each control direct VCC. Further information on the VCCs can be

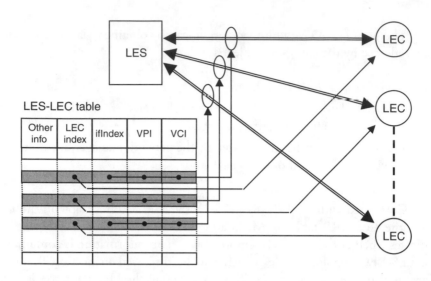

Figure 9.24 Topological mapping using the LES-LEC table.

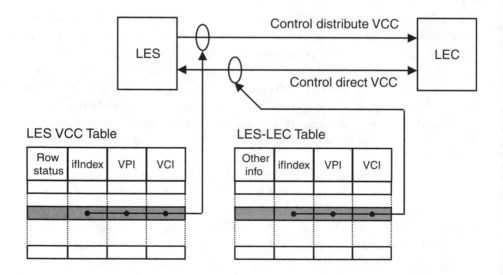

Figure 9.25 LES VCC summary.

obtained from the AToM MIB or other appropriate ATM MIB modules implemented.

9.5.4.2 LES Performance Management

Fault management involves two groups, the LES statistics group and the LES-LEC statistics group. The former provides general statistics of an ELAN as perceived by the LES. The latter is focused on the performance of ARP for each LES-LEC pair.

9.5.4.2.1 LES Statistics Group

This table provides performance and fault counters on a per-LES basis. Since it is an extension to the LES configuration table, this table augments it.

The LES statistics table is similar to the LECS statistics table in the LECS MIB. The table also lists all error types regarding an LES listed in the Table 13 [1]. The counters keep the numbers of events as follows:

- Successful join responses sent out by the LES;
- Version not supported errors;
- Invalid request parameters errors;
- Duplicate LAN destination errors;
- Duplicate ATM address errors;
- Insufficient resources to grant errors;
- Access denied for security reasons errors;
- Invalid LEC ID errors;
- Invalid LAN destination errors;
- Invalid ATM address;
- Malformed ATM ARP requests received;
- Registration failures sent;
- LE_ARP_REQUESTs received;
- LE_ARP_REQUESTs forwarded onto the clients of the LES.

9.5.4.2.2 LES-LEC Statistics Group

This group is designed to monitor the performance of the LE_ARP. It contains all LE_ARP request related counters and error counts on a per-LEC-LES pair basis. Each entry in this table represents an LEC and its ARP counters. This table is an extension to the LES LEC table and thus augments it. The counters in this table keep a track of the numbers of following events for the specified LES-LEC pair:

- Requests received including all control frames as well as LE-ARP requests;
- Requests or responses sent;
- Register requests received;
- UnRegister requests received;
- LE-ARP requests for UNICAST address received;
- LE-ARP requests for MULTICAST and broadcast address received;
- LE-ARP responses received;
- NARP requests received.

9.5.4.3 LES Fault Management Group

This group provides fault management information for managing LESs this agent supports. The fault management mechanism in LES MIB is the same as that of the LECS MIB. For a detailed description, please refer to Section 9.5.4.

9.5.5 Broadcast and Unknown Server MIB

The BUS MIB is designed for the management of LANE BUSs. In particular, this MIB enables network managers to create, destroy, and configure BUSs that the agent manages and to determine the current status and performance statistics of BUSs and topology of the portions of ELANs being served by BUSs. This MIB should be used in conjunction with the ELAN and LES MIBs.

The structure of the MIB is depicted in Figure 9.26, which is almost the same as the LES MIB, except that LES-specific tables have been deleted. These tables include the LES-BUS Table and LES ARP tables. The objects defined in corresponding tables in the LES and the BUS MIBs are similar to each other. Accordingly, only the differences between these two MIBs are given.

9.5.5.1 BUS Configuration Group

This group includes the object busNextId, BUS configuration table, BUS VCC table, and BUS-LEC topology table. The object busNextId provides the network manager with the next available index used to create a BUS instance.

9.5.5.1.1 BUS Configuration Table

This table lists all BUS instances this agent manages. Each entry in this table represents a BUS. The table contains the ATM address and the status objects in exactly the same way as those in the LES configuration table, which was

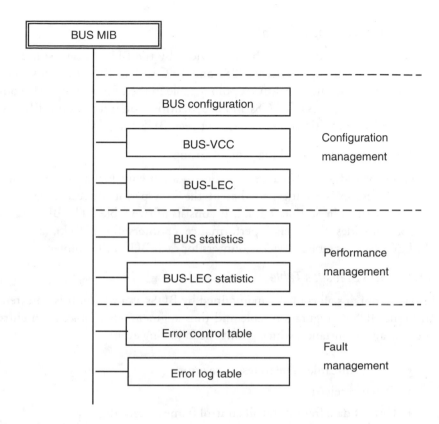

Figure 9.26 Structure of the BUS MIB.

introduced in Section 9.5.4.1.1. However, LES-specific information, such as the ELAN type and frame size, is replaced by a BUS specific object, the max frame age. This is the time-out period for a frame that has been received but not been transmitted by BUS to all relevant multicast send VCCs or multicast forward VCCs. The table also keeps the ELAN name object to indicate the ELAN this BUS is servicing and is used to cross-reference tables defined in the LES MIB.

9.5.5.1.2 BUS VCC Table

This table lists all multicast forward VCCs used by the BUS to forward multicast traffic to the participating LECS. This table is the same as the LECS or the LES VCC table. Please refer to Section 9.5.4.1.2 for details.

9.5.5.1.3　BUS-LEC Topology Table

This table lists the actual LECS being serviced by the BUS. It can be used to determine the mapping between BUSs and LECS. The BUS-LEC topology table is simpler than the LES-LEC topology table in the LES MIB. It only contains the ATM addresses of LECSs and the identification information (ifIndex, VPI and VCI) for multicast send VCCs between BUS and LECS.

9.5.5.2　BUS Performance Management Group

This group provides performance statistics to a network manager. The structure of this group is identical to that of the LES performance management group as described in Section 9.5.4.2. It consists of two tables: The BUS statistics table provides the overall performance counters for a BUS, and the BUS-LEC statistics table provides counters for each BUS-LEC connection.

9.5.5.2.1　BUS Statistics Table

This table contains all counters maintained by BUSs as a whole. It is an extension to the BUS configuration table and thus augments that table. Each entry represents a BUS instance, which count the following events:

- Frames discarded due to resource error;
- Octets received;
- Unicast data frames and all control frames received;
- Multicast frames received;
- Frames dropped due to time out;
- Refused multicast send VCC connection set-up attempts to the BUS;
- Unsuccessful multicast forward VCC connection set-up attempts from the BUS.

9.5.5.2.2　BUS-LEC Statistics Table

This table contains all LEC counters the BUS maintains. It is an extension to the BUS-LEC topology table and thus augments that table. Each entry represents a BUS-LEC pair, which counts multicast, broadcast, and unknown forward requests such as received, forwarded, and discarded, respectively.

9.5.5.3 BUS Fault Management Group

This group provides fault management information for managing a BUS. Again, the mechanism used is exactly the same as that in the LECS fault management group, as discussed in Section 9.5.3.3.3.

References

[1]   ATM Forum, "LAN Emulation Over ATM Specification, Version 1.0," af-lane-0021-000, January 1995.

[2]   ATM Forum, "LAN Emulation Over ATM Version 2 - LUNI Specification," af-lane-0084.000, July 1997.

[3]   ATM Forum, "LAN Emulation Over ATM Specification, Version 1.0 Addendum," af-lane-0050-000, December 1995.

[4]   ATM Forum, LAN Emulation—Client Management Specification, Version 1.0," af-lane-0038-000, September 1995.

[5]   ATM Forum, "LAN Emulation—Server Management Specification, Version 1.0," af-lane-0057-000, March 1996.

[6]   Internet Drafts, Ly, F., M. Noto, A. Smith, K. Tesink, "Definitions of Supplemental Managed Objects for ATM Management," (draft-ietf-AToM MIB-atm2-01.txt).

10

The Role of Proprietary MIB

10.1 Introduction

ATM technology has matured enormously in recent years, as a result of the efforts of various bodies such as ITU-T, IETF, and, especially, the ATM Forum. Formed in 1991 by only a few companies, the ATM Forum now consists of more than 900 member companies and plays a key role in the development and recommendation of interoperability specifications and the promotion of industry cooperation and awareness. The ATM Forum has completed a number of key specifications, including ILMI, signaling, traffic management, PNNI, LANE, MPOA, and network management, which enables truly scalable, interoperable, and manageable ATM networks. Most of these specifications were completed in the last two or three years.

The incorporation of these standards, however, requires much higher performance processors to deliver the high call set-up rates or the low call set-up latencies for the practical operation of ATM internetworking protocols. Moreover, many functions may need to be supported directly by hardware, given the extremely high speed of ATM systems. As a result, few, if any, of the first generation of ATM switches have the capability to implement these complex and demanding new specifications. In fact, just like any other field of technical endeavor, pioneering ATM vendors began with prestandard or proprietary solutions simply because the market has not yet reached the critical mass that allows standards to form. As the ATM industry matures, these proprietary mechanisms must be supplanted by standard mechanisms.

The Cisco *LightStream 1010* (LS1010) is among the first ATM switches designed to fully support the latest ATM Forum specifications while delivering the performance, scalability, robustness, and manageability required for true production ATM deployment. Detailed information on LS1010 can be found in [1].

Part Three of this book is devoted to demonstrating how standard ATM MIBs from various standardization bodies and proprietary vendor-specific MIBs are combined to manage ATM switches by examining LS1010.

10.2   Cisco LightStream 1010

The LS1010 is a 5-Gbps modular switch designed for either workgroup or campus applications, depending on the nature of the interfaces employed. At the center of the switch, there is a single, field-replaceable *ATM switch processor* (ASP) module that supports both the 5-Gbps shared memory, fully nonblocking switch fabric, together with its feature card, and the high-performance RISC processor that provides the central intelligence for the device. The ASP module feature card can also be upgraded in the field, allowing the switch to track changing ATM specifications. Currently, LS1010 supports almost all major ATM standards including UNI 3.0, UNI 3.1, UNI 4.0, ILMI, PNNI Phase 1, IISP, LANE, Soft PVC/PVP, and CES.

The switching fabric of the LS1010 contains 65,536 cells of shared buffer memory that is shared across ports. The cell switching is performed when reading cells from the shared memory pool. This switching fabric, along with advanced buffer allocation mechanisms and policies, minimizes cell loss and allows for flexible service support. The architecture of the shared memory fabric also facilitates easy field upgradability to support future standards or advanced capabilities, because all value-added switch mechanisms are centralized on the field-upgradable feature card on the ASP module.

A full range of interfaces are supported on a series of port adapter modules, each of which supports multiple ports allowing for full utilization of the switch capacity. For example, a LS1010 can support up to 32 SONET ST3c/STM1 155-Mbps ports, up to eight SONET STS12c/STM4c 622-Mbps ports, or up to 192 25-Mbps ATM ports, with a total capacity of 5-Gbps. The available interfaces for LS1010 are as follows:

- SONET STS-3c/STM-1 155 Mbps;
- SONET STS-12 /STM-12 622-Mbps;

- T1/E1 circuit emulation;
- T1/E1 (ATM);
- DSE3/E3;
- 25.6-Mbps ATM.

The LS1010 offers the sophistication and depth of functionality required for true ATM production deployment. Advanced traffic management mechanisms allow for the support of current, bursty, best-effort traffic, while also delivering the QoS guarantees required for the applications of the future. ABR congestion control support allows the LS1010 to slow traffic sources before congestion becomes excessive, while support for the ATM Forum PNNI protocols allows networks of LS1010s to scale to hundreds of nodes while still delivering interoperable, QoS-based routing. Value-added capabilities allow for ATM access lists and load sharing across redundant links. All of this sophistication is hidden, however, by the true standards-based, plug-and-play capabilities of the LS1010, and advanced management functions allow for unprecedented levels of network visibility and control.

ATM management is particularly challenging not only because of the general unfamiliarity of ATM but also because of the relative immaturity of ATM management standards. As a matter of fact, the IETF and the ATM Forum have been working together on defining or refining a series of MIBs to allow for the configuration and monitoring of particular aspects of the operation of ATM switches. In addition to these standard MIBs, vendors also need to implement comprehensive private MIBs to complement and supplement the standards. The LS1010 supports almost all standard ATM MIBs and prestandard extensions. An overview of all standard and Cisco proprietary MIBs will be presented in Section 10.3.

Monitoring of ATM traffic flows poses further difficulties because of the speed of the ATM links. As a result, dedicated hardware is typically required in order to allow for real-time statistics collection. Consequently, the number of statistics gathered per connection and the number of connections for which statistics are to be gathered need to be kept to reasonable bounds. The reason is that gathering such statistics requires many hardware counters to be built into the switching fabric, in addition to ancillary memory for data storage, all of which add significantly to switch cost and complexity. The switching fabric of LS1010 is capable of supporting up to 32,000 point-to-point and 1,985 point-to-multipoint connections. This is equivalent to 1,000 point-to-point connections and a smaller number of point-to-multipoint connections per port, if a switch is configured to support 32 SONET STS-3c/STM-1 155-Mbps ports.

In addition to its SNMP-based management, LS1010 supports standard ATM OAM cell flows since it will allow for rapid detection and notification of failed links and connections. In particular, it supports F4 and F5 ATM OAM segment and end-to-end flows and value-added OAM ping functions for link and connection integrity testing, which will be introduced in Chapter 13.

Moreover, LS1010 offers the capabilities of port snooping and circuit steering, which enables real-time monitoring of the traffic on particular switch connections. Unlike shared medium networking technologies, there is no easy way to monitor traffic through an ATM switch without explicit switch support. With port snooping and circuit steering capabilities, an ATM switch can divert all or some of the connections through a particular port on the switch from which they could then be sent to an external ATM analyzer or an embedded ATM RMON analyzer.

10.3 Proprietary MIB

SNMP is an open framework that has been developed in such a way that standard Internet MIB (i.e., MIB-II) can easily be extended to allow the addition of MIB modules for a variety of network technologies and other network entities. Internet standard MIB extensions are either added into the "mgmt.mib-2" subtree directly (e.g., AToM MIB) or to the transmission subtree under "mgmt.mib-2" (e.g., SONET MIB).

Experimental MIBs being developed by IETF working groups define objects under the *experimental* branch. IETF working groups obtain a number under the experimental branch through coordination with the network management area director and the IANA. When an information module produced by a working group becomes a *standard* information module, then at the very beginning of its entry onto the Internet standards track, the objects are moved under the "mgmt" subtree.

Proprietary MIBs define objects within an organization's subtree located under the *enterprises* branch of the *private* subtree. Administration of this subtree is delegated by the IAB to the IANA for the Internet. An enterprise can request a node under the *enterprises* subtree from IANA to register its management objects. Upon receiving a subtree, the enterprise may, for example, define new MIB objects in this subtree. In addition, the enterprise will usually also register its networking subsystems under this subtree in order to provide an unambiguous identification mechanism for use in management protocols. A list of the enterprise numbers assigned by IANA is available from [2]. Figure 10.1 shows a few nodes in the *enterprises* subtree. The OID for Cisco is 1.3.6.1.4.1.9.

The mechanism of incorporating proprietary MIBs into the SNMP framework makes it possible for vendors to add vendor-specific attributes to standard MIB modules. For example, LS1010 ATM switches that implement the standard MIB-II objects also have a private extension of new objects that apply specifically to Cisco LS1010. In some cases, the vendor-specific ones cover the same functions as the standard ones, as a result of vendor's needs at times when the standard ones were not finished yet. Actually this provides another way of defining new standard MIB modules. New extensions may start out as proprietary MIBs. After testing, a MIB module may be moved from the *enterprises* subtree to the standard portion of the MIB tree. This approach is especially useful for ATM, where the technology is still evolving. In fact, Cisco ATM extensions contain many prestandard objects. A consequence of this, however, is that the vendor-specific ones are often in use a long time after the standard ones are matured.

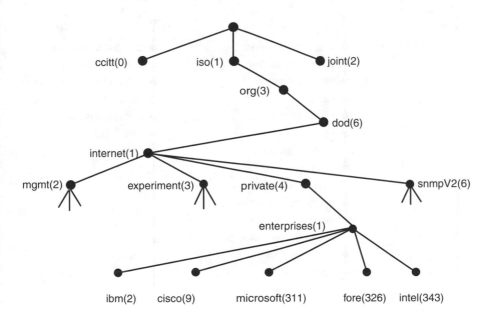

Figure 10.1 Example of the nodes under the enterprises subtree.

10.4   Overview of LS1010 MIBs

Cisco manages it enterprise subtree in much the same way as Internet does. It assigns 16 top-level OIDs according to the Cisco Enterprise Structure of Management Information [1, 3]. Among them, three nodes are of special interest in managing the LS1010 ATM switch, which is depicted in Figure 10.2, along with other Cisco proprietary MIB modules implemented by the Cisco IOS Software Release 11.2.2 for the LS1010 ATM switch.

The roles of the MIB subtrees shown in Figure 10.2 are summarized as follows:

- workgroup(5): Subtree reserved for use by the Workgroup Business Unit in Cisco;

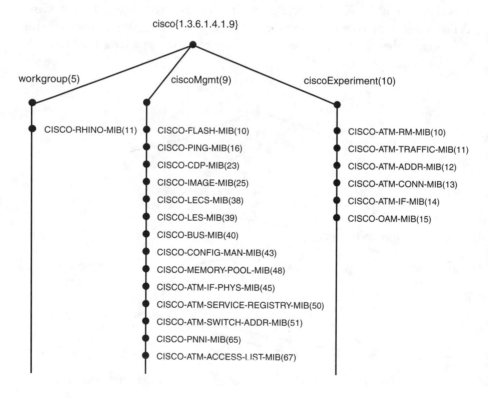

Figure 10.2   Subtrees used for managing LS1010.

- ciscoMgmt(9): Main subtree for new MIB development in Cisco;

- ciscoExperiment(10): Just like the "experimental" sub-tree in the Internet, provides a root object identifier from which experimental MIBs may be temporarily based. MIBs are typically based here if they fall in one of two categories:

  - IETF work-in-process MIBs, which have not been assigned a permanent object identifier by the IANA;

  - Cisco work-in-process, which has not been assigned a permanent OID by the Cisco assigned number authority, typically because the MIB is not ready for deployment.

LS 1010 supports both SNMPv1 and SNMPv2 (party-based) using the bilingual agent mechanism described in Chapter 2. Bilingual support of SNMP gives users optimal flexibility by allowing them to migrate to SNMPv2 on their own timetables. During the period when SNMPv1 and SNMPv2 coexist, LS1010 users will not lose any management functionality. LS1010 software allows customers to configure the version of SNMP between the agent and the manager to the one supported by the management station. Since LS1010 is capable of talking to multiple NMSs with different versions of SNMP protocol, it can be configured to support SNMPv1 or SNMPv2 only, or SNMPv1 and SNMPv2 simultaneously.

There are four ways in which a network manager can access the SNMP agent in a LS1010:

- In-band ATM via Ethernet LEC;

- In-band ATM via RFC 1577 classical IP over ATM client;

- Out-of-band via Ethernet;

- Out-of-band via serial interface.

The management of the LS1010 involves a number of MIBs, including ATM-specific MIBs and non-ATM-specific MIBs, both consisting of standard MIBs and Cisco enterprise MIBs. These MIBs will be briefly introduced in terms of their respective roles. Detailed description of how the MIBs are used together to manage ATM connections, traffic, and OAM functions will be provided in Chapters 11–13.

10.4.1   General Management

Although LS1010 is an ATM switch, it is also required to fulfill a set of common management tasks such as the support of standard Internet MIB (i.e., MIB-II and its extension IF-MIB) as well as SNMP MIBs. In addition, the hardware and software of LS1010 also need to be managed, which involves a set of Cisco-specific MIBs. The MIBs supported for general management are summarized as follows:

- Standard MIBs:
  - RFC 1213: MIB-II;
  - RFC 1573: IF MIB;
  - RFC 1450: Party-based SNMP v2 MIB;
  - RFC 1447: Party-based SNMP v2 Party MIB;
  - RFC 1659: RS232 MIB.
- Cisco-specific MIBs:
  - CISCO-FLASH-MIB: Flash Memory MIB; Flash memory is used to upgrade LS1010 software to facilitate a switch to support future functions (i.e., new ATM standards);
  - CISCO-CDP-MIB: The Cisco Discovery Protocol MIB;
  - CISCO-IMAGE-MIB: Allows a NMS to query the capabilities of the software running on a Cisco device;
  - CISCO-CONFIG-MAN-MIB: The Cisco configuration management MIB, which keeps configuration data that exists in various locations, observes the overall state of the system configuration, and records the time and source of changes;
  - CISCO-MEMORY-POOL-MIB: The Cisco MIB for monitoring memory pools such as the processor memory;
  - CISCO-RHINO-MIB: The chassis MIB for the LS1010 ATM switch.

Note that there are some other old Cisco MIBs that are also supported by the LS1010. However, they are out of the scope of this book.

10.4.2   ATM Layer

ATM layer management encompasses the AToM MIB and a few Cisco extensions (depicted in Figure 10.3), the role of each which is given in Table 10.1.

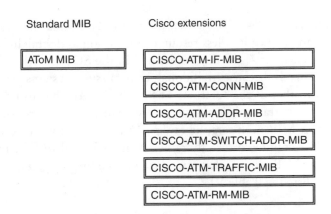

Standard MIB   Cisco extensions

| ATOM MIB |

| CISCO-ATM-IF-MIB |
| CISCO-ATM-CONN-MIB |
| CISCO-ATM-ADDR-MIB |
| CISCO-ATM-SWITCH-ADDR-MIB |
| CISCO-ATM-TRAFFIC-MIB |
| CISCO-ATM-RM-MIB |

Figure 10.3 ATM layer management.

Table 10.1
Cisco ATM MIB Extensions

Cisco Extension	Description
CISCO-ATM-ADDR-MIB	A list of the valid calling party addresses for a UNI on a per-interface basis
CISCO-ATM-CONN-MIB	An extension to VPL/VCL table defined in RFC 1695 for ATM switch connection management for SVP, SVC
CISCO-ATM-IF-MIB	An extension to RFC 1695 ATM interface table
CISCO-ATM-RM-MIB	Cisco ATM switch resource management MIB to manage switch resources including shared memory cell buffers and interfaces
CISCO-ATM-SWITCH-ADDR-MIB	A list of all ATM addresses for the switch
CISCO-ATM-TRAFFIC-MIB	An extension to traffic OIDs and variables defined in RFC 1695.

10.4.3 Physical Layer

LS1010 does not use any standard physical layer MIBs; instead, it employs a subset of objects from the SONET and DS3 MIB into its own

CISCO-ATM-IF-PHYS-MIB. In fact, other ATM vendors such as Fore Systems do the same, since it is less resource-consuming and straightforward. As introduced in Chapter 7, the objects defined in the SONET MIB are derived parameters. This is because traditional transmission systems are managed differently from that of the Internet. The SONET MIB has to be defined in the way in which transmission systems are managed. In contrast, CISCO-ATM-IF-PHYS-MIB monitors the raw parameters directly. For example, the parity errors are countered directly rather than converted to ESs. Figure 10.4 summarizes the MIB module that is used to manage the physical layer. The role of this MIB is listed in Table 10.2.

10.4.4 ILMI

Besides the standard ATM Forum's ILMI MIBs that support the ILMI 4.0 specification, the management of ILMI includes the CISCO-ATM-SERVICE-REGISTRY-MIB. As discussed in Chapter 5, the managed information contained in the ILMI MIBs is not directly available to NMSs unless a proxy is implemented. Moreover, the objects in the ATM Forum's service registration MIB cannot be modified, as they are either not-accessible or read-only. The Cisco extension includes almost identical objects as defined in forum's MIB; however, this MIB is available to NMSs and the access level is changed to "read-create." This allows an NMS to monitor and configure the

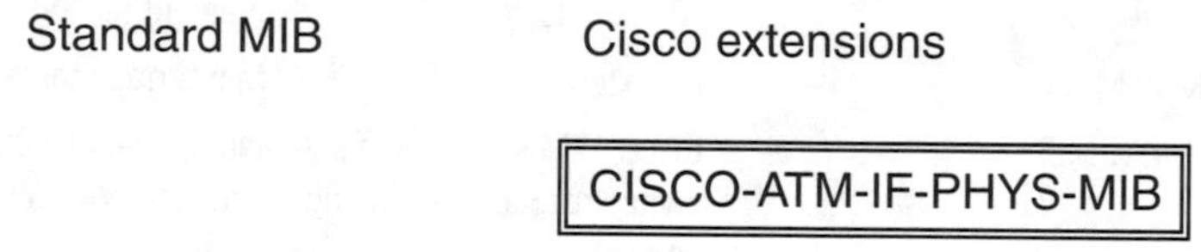

Standard MIB   Cisco extensions

> CISCO-ATM-IF-PHYS-MIB

Figure 10.4 Physical layer management.

Table 10.2
Physical Layer Management

Cisco Extension	Description
CISCO-ATM-IF-PHYS-MIB	A subset of SONET and DS3 MIB containing basic status and statistics for the physical layer of ATM interfaces

information that an ATM switch makes available via the ILMI's service registry table. Figure 10.5 shows the MIBs that are relevant to ILMI functions. Table 10.3 is a brief description of the Cisco ILMI extension.

10.4.5 LANE

As of this writing, LS1010 does not support LANE management. However, from the MIBs that are available from the Cisco, it is anticipated that LS1010 LANE management will involve at least the standard LEC MIB for client management and three Cisco proprietary MIB modules for service management.

Cisco's services MIBs are much smaller than the ATM Forum's counterparts, which suggests a possible reason why Cisco uses proprietary MIBs (i.e., the complexity of implementing ATM Forum's LANE service MIBs). ATM Forum's MIBs specify sophisticated management information for flexible and interoperable operation. Consequently, they are difficult to implement and

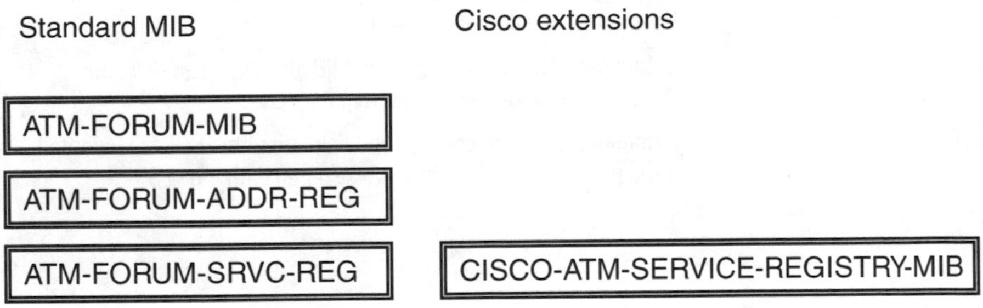

Standard MIB   Cisco extensions

| ATM-FORUM-MIB |

| ATM-FORUM-ADDR-REG |

| ATM-FORUM-SRVC-REG | | CISCO-ATM-SERVICE-REGISTRY-MIB |

Figure 10.5 ILMI management.

Table 10.3
Cisco ILMI Extension

Cisco Extension	Description
CISCO-ATM-SERVICE-REGISTRY-MIB	Contains almost identical objects as defined in the ILMI's service registry table to allow an NMS to monitor and configure the information that an ATM switch makes available via that MIB.

require more resources to operate. For example, the ATM Forum's LANE BUS MIB contains 10 groups, whereas the CISCO-BUS-MIB contains just two groups, supporting only VCC and configuration management, as well as simple statistics. As a result, the latter does not include the full functionality for fault management and performance management as defined in the ATM Forum's MIB, nor does the CISCO-LECS-MIB or CISCO-LES-MIB. The functions of the Cisco service MIBs are briefly described in Table 10.4. Figure 10.6 depicts the MIBs that are employed for the management of LANE.

Table 10.4
Cisco Proprietary LANE Service MIBs

Cisco Extension	Description
CISCO-LECS-MIB	Creating, configuring, and monitoring the LECS as well as entering/modifying data within the LECS database
CISCO-LES-MIB	Creating/monitoring the LES; this MIB also allows the monitoring of the LANE clients as perceived from the server
CISCO-BUS-MIB	Monitoring the BUS; creation and deletion of the BUS is facilitated through the LES MIB as the BUS is shared with the LES

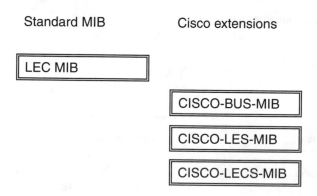

Figure 10.6 LANE management.

10.4.6 PNNI

The PNNI is such a complicated protocol that its implementation is anticipated to be gradual and piecemeal. LS1010 implements only a portion of the PNNI phase 1 specification. Current version of the LS1010 software supports 10 groups of objects, including the base group, node table, NodeTimerTable, scope mapping table, summary table, IfTable, link table, neighboring peer table, neighboring peer port table, and route address table. There is also an extension, CISCO-PNNI-MIB, containing Cisco-specific objects to support Cisco-specific functions, such as background routing configuration, route optimization, and NSAP to E.164 ATM address translation. PNNI-specific MIBs supported in LS1010 are summarized in Figure 10.7. Table 10.5 describes the Cisco extension for PNNI.

10.4.7 Remote Network Monitoring

Remote monitoring MIBs add high-level SNMP traffic monitoring capability to the LS1010 system. The newly completed ATM Forum's ATM RMON MIB extends traditional RMON data sets such as flow statistics, host, and traffic matrix to ATM networks. The ATM RMON MIB can be enabled on a

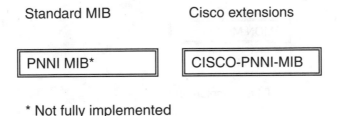

Figure 10.7 PNNI management.

Table 10.5
Cisco Extension for PNNI

Cisco Extension	Description
CISCO-PNNI-MIB	The MIB module for managing Cisco-specific extensions to the ATM Forum PNNI MIB

per-interface basis. LS1010 implements the alarm and event group of the
RMON MIB and the ATM RMON MIB. Figure 10.8 shows the MIBs
employed for remote monitoring. Table 10.6 briefly lists the roles of these
MIBs.

10.4.8 Accounting

Accounting management permits network administrators to collect accounting data based on ATM bandwidth/resource usage per connection within
the ATM network, which provides a basis for optimum use of an enterprise's networked resources. LS1010 accounting management involves two
MIB modules that are under development by the IETF AToM working
group, ACCOUNTING-CONTROL-MIB and ATM-ACCOUNTING-
INFORMATION- MIB; these modules keep key, per-connection information
storage for subsequent use by accounting and billing applications. Chapter 14
briefly introduces these two MIBs that are depicted in Figure 10.9. Their functions are described in Table 10.7.

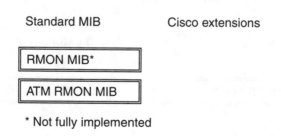

Figure 10.8 RMON management.

Table 10.6
RMON MIBs for LS 1010

MIB Name	Description
RMON MIB	IETF RMON MIB
ATM-RMON-MIB	ATM remote monitoring MIB from the ATM Forum

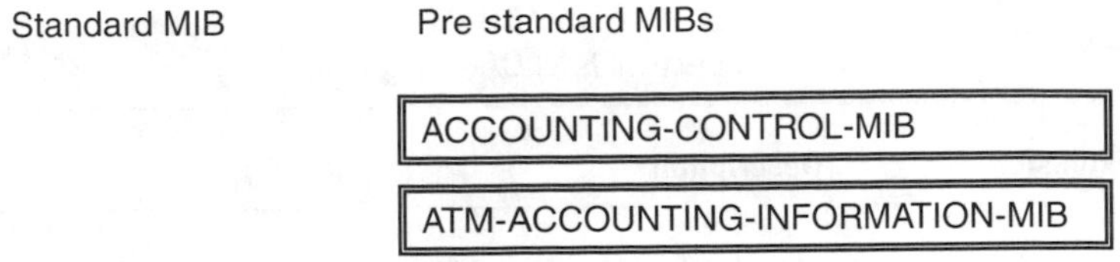

Standard MIB   Pre standard MIBs

| ACCOUNTING-CONTROL-MIB |
| ATM-ACCOUNTING-INFORMATION-MIB |

Figure 10.9 Accounting management.

Table 10.7
Accounting MIBs

MIB Name	Description
ACCOUNTING-CONTROL-MIB	Managing the collection and storage of accounting information for connections in a connection-oriented network such as ATM
ATM-ACCOUNTING-INFORMATION-MIB	Identifying items of accounting information that are applicable to ATM connections

10.4.9   OAM Ping

OAM Ping is a value-added function offered by Cisco. It is implemented by including an IP or ATM address within an OAM cell. This allows the network administrator to verify link or connection integrity at any intermediate point, thereby facilitating network troubleshooting. To invoke this function, the CISCO-OAM-MIB must be used to permit OAM loop-back ping on ATM connections. A detailed description can be found in Chapter 13. Figure 10.10 shows the MIB that is used for OAM Ping. Table 10.8 summarizes the role of the MIB.

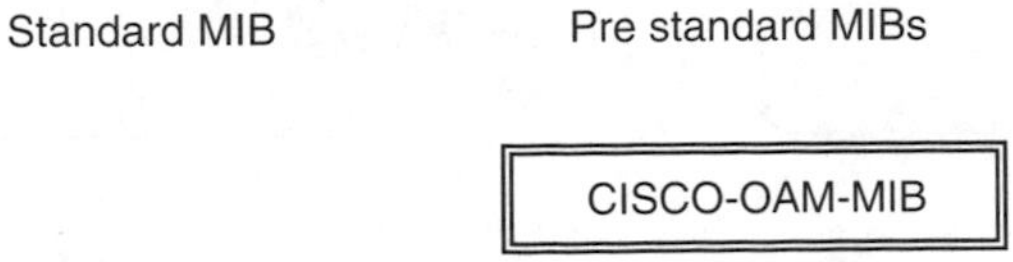

Standard MIB   Pre standard MIBs

| CISCO-OAM-MIB |

Figure 10.10 MIBs for OAM Ping.

Table 10.8
Cisco MIBs for OAM Ping

Cisco Extension	Description
CISCO-OAM-MIB	MIB module for invoking OAM loop-back ping on ATM connections

10.4.10   Access Control

Access control is a mechanism to protect network resources from unwanted or malicious attacks. It is a widely used security mechanism in TCP/IP-based Internet, where it is known as Firewall. The CISCO-ATM-ACCESS-LIST-MIB is capable of configuring and controlling access control filters in an ATM switch. Access control is in the security management domain. As usual, Figure 10.11 shows the MIB that is used for access control management; and Table 10.9 describes its role.

Standard MIB   Cisco Extensions

CISCO-ATM-ACCESS-LIST-MIB

Figure 10.11 Access control management.

Table 10.9
Cisco Access Control MIB

Cisco Extension	Description
CISCO-ATM-ACCESS-LIST-MIB	An MIB for configuration and control of access control filters in an ATM switch

References

[1]   Cisco Systems Inc., http://www.cisco.com/.

[2]   IANA, IANA-Enterprise Numbers, ftp.isi.edu/in-notes/iana/assignments/enterprise-numbers.

[3]   FTP site for Cisco MIBs, ftp://ftp.cisco.com/pub/mibs/.

11

ATM Connection Management

11.1 Introduction

Connection management is essential to managing ATM, given the connection-oriented nature of ATM networks. Four types of connections have been standardized by the ATM Forum [1–4]:

1. Permanent virtual connections, including PVPC and PVCC, are established administratively (i.e., by network management). This type of virtual connection is offered on a provisional basis. Network management is responsible for the set-up, tear down, and monitoring of a permanent connection.

2. SVCs, including SVPC and SVCC, are set up in real time on demand via call set-up signaling procedures across the UNI. The major objective of managing this type of connection is to monitor the operation of switched connections because they consume resources of ATM switches. However, a network manager should be able to tear down an established connection in the event of abnormality.

3. Soft virtual connections, including soft VPC and soft VCC, are connections of which portions are switched, while other portions are permanent. Therefore, a soft connection is established cooperatively by network management and signaling. It is the network management rather than the user that should provide information necessary to establish the SVC within the network by signaling. Therefore, manag-

ing a soft connection requires more management information than managing a permanent or a switched connection alone.

4. Logical ports, also known as VP tunneling, are VPs that are used as a trunk connection (most likely between devices that are not physically adjacent), providing for multiplexing and demultiplexing of VCs on the VP. Again, more information is needed for managing a logical port than an ordinary ATM interface.

Currently, ATM layer management is carried out at three levels described as follows.

1. The ATM interface level manages the ATM interface within a managed device as a whole and provides aggregated management information of configuration parameters, status, and performance statistics per ATM interface. Note that ATM interface management only has local significance, dealing with the local interface, or the interface between the adjacent physical or logical interface.

2. VP level provides management information per VP. The end-to-end management of a VPC is achieved by managing the VPLs of which the VPCs consist.

3. VC level management is similar to that of the VP level.

Both the ATM Forum's ILMI MIB [5] and the IETF's RFC 1695 AToM MIB [6] apply to all three levels, but their roles are different as already discussed in Chapters 5 and 6. The ILMI MIB is focused on local link management, while the AToM MIB is more end-to-end-oriented, although it can be used for both link management and end-to-end management. This chapter focuses on virtual connection management, including the VP and VC levels.

The AToM MIB is targeted primarily for permanent virtual connection management, although SVCs are also represented by the management information in the MIB. Full management of SVCs requires additional capabilities that are beyond its scope, let alone the management of soft virtual connections or logical ports. In fact, when RFC 1695 was published, many of the ATM Forum's major specifications, such as PNNI, traffic management 4.0, and UNI 4.0, had not been standardized. As a result, the current AToM MIB is in line with the UNI 3.0/3.1 [1, 2] specifications. However, the latest ILMI MIB [5], that is ILMI MIB 4.0, has been updated to reflect the new standards.

There is no doubt that the discrepancies in standards complicate the task of ATM connection management. Furthermore, there are some areas that have not been addressed by any standard at all. Some of them, such as OAM, are generic, whereas others are vendor-specific. In this chapter, we will discuss how Cisco deals with these problems.

11.2 LS1010 Connection Management MIB

The connection management within LS1010 mainly involves the RFC 1695 AToM MIB and the CISCO-ATM-CONN-MIB [7], which is shown in Figure 11.1. This Cisco MIB extends the VPL/VCL table in the AToM MIB in order to manage the four types of connections as listed above and to support OAM functionality. Besides, it contains additional address information for managing SVCs.

Figure 11.2 depicts the categories of management information that are used to manage LS1010 virtual connections. In addition to the AToM MIB and the CISCO-CONN-MIB, the ILMI MIB is included in the figure in order to show different roles of all relevant MIBs. The role of the ILMI MIB is limited to local interface management; thus no cross-connect information is specified. With ILMI, the management information of the adjacent ATM devices can be obtained, while the AToM MIB and the Cisco extension only manage the ATM device itself. However, objects in the ILMI MIB are not accessible from an NMS.

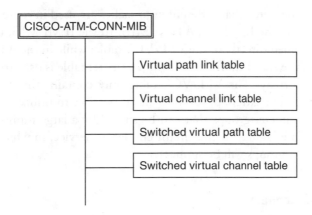

Figure 11.1 Structure of the CISCO-ATM-CONN-MIB.

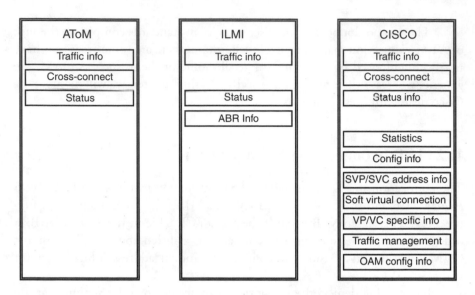

Figure 11.2 Management information used in connection management.

11.2.1   Traffic Information

Traffic information defined in the AToM MIB does not reflect the latest development in the ATM traffic model. To solve this problem, Cisco defines another MIB, namely the CISCO-ATM-TRAFFIC-MIB, to extend the traffic OIDs and variables defined in RFC1695, which will be discussed in detail in Chapter 12.

It should be noted that different approaches are used in representing traffic information in the ILMI and AToM MIB. In the ILMI MIB, traffic information is contained directly in the VPL/VCL table; while in the AToM MIB, a separate table, the ATM traffic descriptor parameter table is used to store traffic information centrally. The VPL/VCL table only contains the indexes to the parameter table. The latter approach makes it possible to utilize a set of predefined traffic parameter groups that can be shared by a large number of virtual connections. One example is the circuit emulation service, in which only a few traffic parameter groups will be supported.

11.2.2   Cross-Connect

There are two ways in which the cross-connect information of virtual connections can be represented, using cross-connect Table in the AToM MIB or using proprietary objects defined in the CISCO-ATM-CONN-MIB, as depicted in Figure 11.3.

Cross-connect information is expressed in the AToM MIB by objects in two tables, the VPL/VCL table and the VP/VC cross-connect table, as explained in Chapter 6. The former contains a cross-connect index to uniquely identify the connection. The latter, on the other hand, provides status information about the connection. This mechanism can be used to support point-to-point, point-to-multipoint, and multipoint-to-multipoint connections. All associated virtual links have the same value of cross-connect identifier, and all their cross-connections are identified by entries in the cross-connect table for which cross-connect index has the same value. However, there is no information about the attribute of virtual links consisting of a point-to-multipoint connection (i.e., root or leaf).

The CISCO-ATM-CONN-MIB, in contrast, defines two sets of virtual link indexes (ifIndex, VPI/VCI), the cross-connect indexes and the next leaf indexes. The former identifies the virtual link to which this virtual link is cross-connected. Their usage to support a point-to-point connection is depicted in Figure 11.4.

The latter is only used when the connection is point-to-multipoint, pointing to the next leaf on the multicast chain, as shown in Figure 11.5. In this case, the meaning of the former set of indexes varies according to the topological position of the virtual link. It identifies the root virtual link for a leaf virtual link entry in the multicast chain, or it is the first leaf for a root virtual link entry. A "0" in the figure means that the pointer is null.

When LS1010s were first released, they actually employed the second approach—i.e., they did not implement the VP/VC cross-connect table in the AToM MIB.

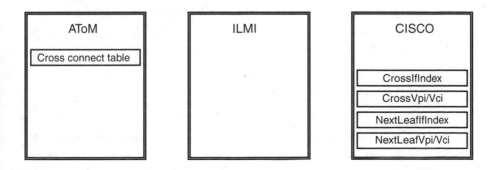

Figure 11.3 Objects for cross-connect.

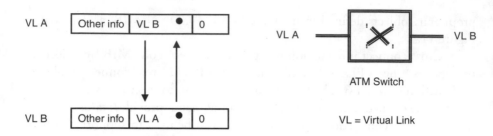

Figure 11.4 Point-to-point connection.

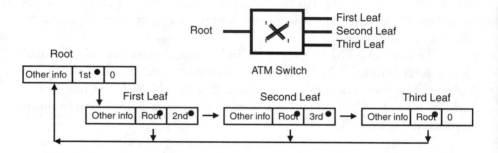

Figure 11.5 Point-to-multipoint connection.

11.2.3 Status Information and Statistics

The CISCO-ATM-CONN-MIB defines two additional status objects:

- ConnState: Connection state of this virtual link (i.e., setup(1), release(2), notInstalled(3), down(4), up(5)). In contrast, the connection status objects only have three states, up(1), down(2), or unknown(3);

- InstallTime: A time stamp derived from sysUpTime when this virtual link is created.

The Cisco MIB also adds two objects to collect the numbers of input and output cells per virtual connection. The objects that provide status and statistics information for ATM connections are shown in Figure 11.6.

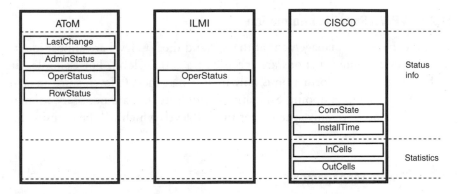

Figure 11.6 Status information and statistics.

11.2.4 Configuration Information

To identify the type of virtual connection, the CISCO-ATM-CONN-MIB specifies three objects as follows:

- Cast Type: Multicast mode (i.e., point-to-point, point-to-multipoint root, point-to-multipoint leaf). Note that LS1010 does not support multipoint-to-multi-point connections directly.

- Span type: Indicating whether the link is a transit one or a terminating one.

- Configuration type: Identifying whether the link is a permanent, switched, or soft virtual connection.

With the extension of these objects to the AToM MIB, an ATM connection can be described much more accurately. For example, the Cast Type object helps to clarify whether a virtual link is root or a leaf when the cross-connect tables in the MIB are used.

11.2.5 Soft Virtual Connection

Managing soft virtual connections requires additional information that will be investigated in Section 11.4.

11.2.6   VP/VC Specific Information

Virtual connection management at the VP and the VC level are almost identical. However, some functions are specific at a certain level. For example, the AAL configuration information is only included at the VC level, since an AAL terminates at a VC endpoint. Similarly, the objects that are necessary to manage a logical port are only specified at the VP level, which will be examined in Section 11.5.

11.2.7   Traffic Management

These include configuration parameters, status, and statistics for per-virtual connection, which will be introduced in Chapter 12.

OAM Configuration Information

These objects are defined to configure the F4/F5 OAM flow, which will be detailed in Chapter 13.

11.3   Permanent Virtual Connection/SVC Management

11.3.1   Permanent Virtual Connection Management

Permanent virtual connection management is illustrated in Figure 11.7, in which a number of object groups are involved. The AToM virtual link table together with the traffic descriptor parameter table jointly provide information about the virtual links of an ATM switch, including traffic characteristics of the links. In addition, the Cisco configuration and status objects offer more detailed information regarding the links. The switching fabric that interconnects virtual links is described by the AToM cross-connect table (see Chapter 6 for details) or by the Cisco cross-connect objects as mentioned earlier in Section 11.2.

11.3.2   SVC Management

SVC management in LS1010 encompasses two additional tables, the switched VP link address table and the switched VC link address table. The structure of the two tables is the same, both containing the virtual link identification information, an object to indicate the direction of the link, and an ATM address of

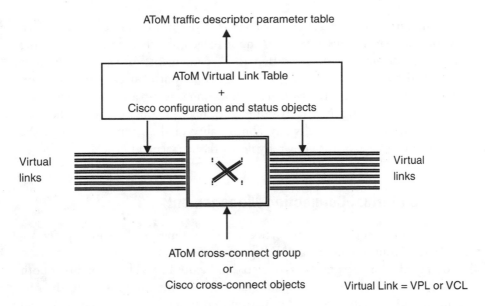

Figure 11.7 Permanent virtual connection management.

the connection depending on the direction. Figure 11.8 depicts the usage of the table. An SVC is established between the source ATM end system and the destination end-system through the ATM switch. For a point-to-point connection, two virtual links will be used. The entry for the link connected to the source end system should set the direction as *calling side* and the address field as the source end-system address, while at the destination side, the entry should contain the destination's ATM address and have the direction set as *called side*.

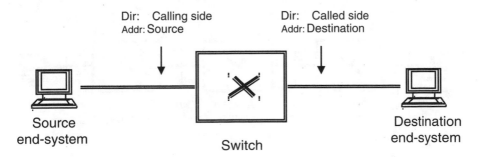

Figure 11.8 Switched virtual link address information.

Although the management of an SVC does not differ very much from that of a permanent one in terms of the managed objects used in LS1010, the ways to implement them are different. A permanent virtual connection is configured by a network manager that has full control over the connection establishment. In contrast, a switched connection is set up and torn down automatically via signaling. Normally, a network manager only monitors the connection, except when there is any abnormality in the connection when the connection needs to be torn down by network management.

11.4 Soft Virtual Connection Management

The soft permanent virtual connection provides a means of setting up permanent virtual connections using PNNI procedures. A soft permanent connection is composed of three segments as depicted in Figure 11.9. The segments at both ends—the one from the source end system to the ingress switch and the one from the egress switch to the destination end system—are permanent virtual connections set up by network management, whereas the central ones are established using PNNI procedures.

The switched portion of a soft virtual connection is established and released between the two *network interfaces* (NIs), which in turn become the endpoints of the soft connection. The endpoints are identified by assigning unique ATM addresses to the corresponding NIs.

One of the two endpoints of a connection, the source endpoint, *owns* the connection and is responsible for establishing and releasing the connection, which is referred to as the calling endpoint. If the switched portion of the connection gets disconnected because of switching system or link failure, it is also

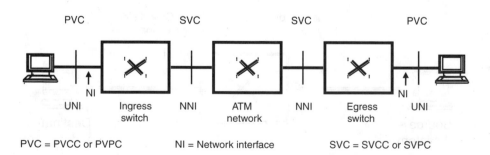

Figure 11.9 Soft virtual connection.

the responsibility of the calling endpoint to try to re-establish the connection. Frequency of re-establishment is an implementation option.

The ingress switch uses PNNI procedures to establish the connection to the egress switch using the ATM addresses of the endpoints of a soft connection as well as information about the VPI/VCI values to be used at the two endpoints by the NMS. The destination switch terminates the connection on the pre-established virtual connection, rather than forwarding an UNI signaling request to the destination end system.

Soft virtual connection establishment offers significant advantages over hop-by-hop permanent virtual connection configuration. Within the network, no manual provisioning is required, thereby simplifying network management. It is a much more convenient and reliable way to set up permanent virtual connections across an ATM network that supports signaling. This also allows permanent connections to be set up with a specific QoS using the PNNI procedures.

Cisco defines additional objects for managing soft virtual connections in the CISCO-ATM-IF-MIB and the CISCO-ATM-CONN-MIB, which is depicted in Figure 11.10.

The SoftVcDestAddress is the endpoint identifier with which this ATM NI can be uniquely identified. The object is defined in the CISCO-ATM-IF-MIB. Other local virtual connection identification information such as VCI/VPI is implicitly provided, because each entry in the CISCO-ATM-CONN MIB is identified by the VCI/VPI object(s) in addition to the ifIndex.

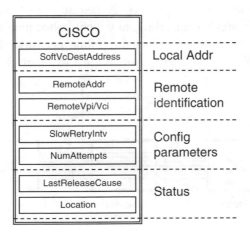

Figure 11.10 Additional objects for managing soft virtual connections.

Similarly, when this switching system wants to set up a soft connection, the identification information is also needed. This includes the endpoint identifier for the remote interface (i.e., the remote NI address) and the VPI or VPI and VCI values to identify the remote virtual connection.

The configuration parameters are defined for the calling endpoint to set up or recover a lost connection due to link or switching system failure. Both the soft PVP/PVC slow retry interval and the number of retries made to install this soft PVC connection are specified.

Finally, two read-only objects are included to reflect the status of a soft virtual connection. The LastReleaseCause object reports the value of the cause field of the cause information element in the last release signaling message received for this connection. The location object indicates the calling or called side of a soft connection.

11.5 Logical Port Management

Logical port, also known as VP tunneling, is a technique to tunnel a bundle of VCCs over a VPC using the VPC service available in ATM Forum UNI 4.0 signaling. This technique is useful in conveying ATM signaling information across the public network, allowing two private ATM networks to be linked using a VP in which the public network transparently trunks the entire collection of VCs in the VP between the two sites. The key advantage is that switched VCCs can be set up between the two private networks, while the requirement for the intermediate public network is minimized—that is, only PVPC. Figure 11.11 illustrates VP tunneling, in which dashed lines denoting signaling VCCs. When a VP is carried across the public network, the order of VCLs within the VP remains unchanged, though the VPLI may change.

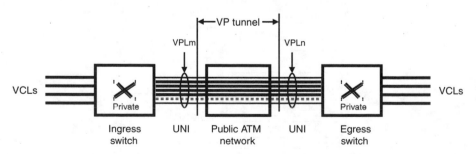

Figure 11.11 VP tunneling.

The logical port is modeled as a logical interface in the MIB-II interfaces table. The interface type "ATM Logical Port" (ifType=80) is defined to allow the representation of a logical port.

Two objects are defined in the Cisco ATM VPL table in the CISCO-ATM-CONN-MIB for managing a logical port:

- LogicalPortDef: Indicates whether the VPC at this VPL interface is an ATM logical port interface;
- LogicalPortIndex: ifTable index of the ATM logical port interface associated with this VPL.

If the VPC is a logical port, then this VPL is associated with an entry of an ATM logical port in the MIB-II interfaces table from which more information is available. The objects that are involved in managing logical ports are summarized in Figure 11.12.

Note that the "ATM Logical Port" interface is more of a logical port, compared with an interface of type "ATM," which is more of a physical port that provides for the transport of both VP and VC connections between adjacent devices.

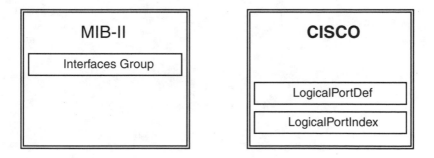

Figure 11.12 Additional objects for logical port management.

References

[1] ATM Forum, "ATM User-Network Interface Specification Version 3.0," September 1993.

[2] ATM Forum, "ATM User-Network Interface Specification, Version 3.1," September 1994.

[3]   ATM Forum, "Private Network-Network Interface Specification Version 1.0 Addendum (Soft PVC MIB)," af-pnni-0066.000, September 1996.

[4]   ATM Forum, "ATM User-Network Interface Signaling Specification Version 4.0," af-sig-0061.000, July 1996.

[5]   ATM Forum, "Integrated Local Management Interface (ILMI) Specification Version 4.0," af-ilmi-0065.000, September 1996.

[6]   Ahmed, M., and K. Tesink, Internet Engineering Task Force, RFC 1695, "Definitions of Managed Objects for ATM Management, Version 8.0 using SMIv2," August 1994.

[7]   Cisco Systems Inc., http://www.cisco.com/.

12

Traffic Management

12.1 Introduction

One of the most important advantages of ATM over other networking technologies is its ability to provide QoS guarantees. ATM traffic management is the key to ensure such guarantees, enabling optimum use of network resources to provide appropriate differentiated QoS for network applications.

ATM networks are used to carry traffic generated by a wide variety of applications. This traffic has different traffic characteristics and thus imposes different requirements on the network over which the traffic is transmitted. ATM traffic management must meet the challenge of balancing high QoS with maximum network utilization. Just like packet switching, cell-based ATM technology is capable of statistically multiplexing different traffic over shared network connections, which results in higher utilization of expensive network resources. Consequently, however, ATM layer congestion may develop because of unpredictable statistical fluctuation of traffic flows and/or fault conditions within the network.

To make things worse, congestion can quickly escalate as higher level protocols request retransmission of lost packets. A network can quickly experience a state of congestion as increasing levels of cell loss generate more and more packet retransmission requests. Preventing congestion or, if this is impossible, controlling congestion so as to maintain a high level of service is a key criteria for traffic management.

ATM traffic management cannot rely on service-specific AAL protocols or on application-specific higher layer protocols. However, protocol layers above the ATM layer may make use of information that may be provided by the ATM layer to improve the utility those protocols can derive from the network.

12.2 ATM Service Categories

The enforcement of QoS through traffic management relies on the definition of service categories. Also requiring definition are parameters by which traffic behavior can be measured to ensure that it is conforming with the QoS characteristics of each category.

The ATM Forum currently defines five service categories [1], which cover the spectrum of potential applications using ATM:

- *Constant bit rate* (CBR);
- *Real-time variable bit rate* (rt-VBR);
- *Non-real-time variable bit rate* (nrt-VBR);
- ABR;
- *Unspecified bit rate* (UBR).

12.2.1 Constant Bit Rate (CBR)

The CBR service category is used by connections that request a fixed (static) amount of bandwidth, characterized by a PCR value that is continuously available during the connection lifetime. The source may emit cells at or below the PCR at any time and for any duration (or may be silent).

This category is intended for real-time applications—i.e. those requiring tightly constrained *cell transfer delay* (CTD) and *cell delay variation* (CDV)—but is not restricted to these applications. It would be appropriate for voice and video applications, as well as for CES.

The basic commitment made by the network is that once the connection is established, the negotiated QoS is assured to all cells conforming to the relevant conformance tests. It is assumed that cells that are delayed beyond the value specified by CTD may be of significantly less value to the application.

12.2.2 Real-Time Variable Bit Rate (rt-VBR)

The rt-VBR service category is intended for time-sensitive applications—i.e. those requiring tightly constrained delay and delay variation, as would be appropriate for voice and video applications. Sources are expected to transmit at a rate that varies with time. Equivalently, the source can be described as *bursty*.

Traffic parameters associated with this service category are PCR, *sustainable cell rate* (SCR) and *maximum burst size* (MBS). Cells that are delayed beyond the value specified by CTD are assumed to be of significantly less value to the application. Real-time VBR service may support statistical multiplexing of real-time sources.

12.2.3 Non-Real-Time Variable Bit Rate (nrt-VBR)

The nrt-VBR service category is intended for applications that have bursty traffic characteristics and do not have tight constraints on delay and delay variation. As for rt-VBR, traffic parameters are PCR, SCR, and MBS. For those cells that are transferred within the traffic contract, the application expects a low CLR. No delay bounds are associated with this category. Nrt-VBR service may support statistical multiplexing of connections.

12.2.4 Available Bit Rate (ABR)

Most of today's ATM applications run over traditional data communication protocols such as TCP/IP, which, in turn, are overlaid over the ATM network. These protocols are sensitive to packet loss but can tolerate variations in delay. In order to reduce loss, there is a need for the network to inform the user of impending congestion with feedback to the sender. The ABR service is intended to implement this feedback.

This service is designed to support applications that cannot effectively characterize their traffic behavior at connection establishment but can adapt their traffic following a flow control protocol. The ABR service makes the promise that if the source behaves in a certain way in response to flow control feedback, then no packets will be intentionally dropped. The initial focus of ABR is the support of LANE connections.

To meet the "vague requirements" for throughput on the establishment of an ABR connection, the end system shall specify a maximum required bandwidth and a minimum usable bandwidth. These are designated as the PCR and the *minimum cell rate* (MCR), respectively. The MCR may be specified as zero. The bandwidth made available from the network may vary, as it is the sum of an MCR and a variable cell rate that results from sharing the available capacity among all the active ABR connections via a defined and fair policy. A flow control mechanism that supports several types of feedback to control the source rate is defined. In particular, a closed-loop feedback control protocol using *resource management* (RM) cells has been specified in a rate-based framework.

No specific QoS parameter is negotiated with the ABR; however, it is expected that an end system that adapts its traffic in accordance with the feedback will experience a low CLR and obtain a fair share of the available bandwidth according to a network-specific allocation policy. CDV is not controlled in this service, although admitted cells are not delayed unnecessarily.

12.2.5 Unspecified Bit Rate (UBR)

The UBR is a "best effort" service—i.e., "send and pray." The UBR service offers no QoS guarantee and does not require any prior knowledge of traffic characteristics. With this service the user is willing to tolerate whatever capacity and cell loss the network can provide at the instant the cell goes through the network. Therefore, UBR is ideally suited to low-cost and low-priority services, such as electronic mail or low-tariff file transfers. UBR service supports a high degree of statistical multiplexing among sources.

A summary of ATM service categories is provided by Table 12.1.

Table 12.1
ATM Service Category Summary [1]

Attribute		Service Category				
		Real-Time		Non-Real-Time		
		CBR	rt-VBR	nrt-VBR	ABR	UBR
Traffic parameters	PCR, CDVT	√	√	√	√	√
	SCR, MBS, CDVT	N/A	√	√	N/A	N/A
	MCR	N/A	N/A	N/A	√	N/A
QoS parameters	Peak-to-peak CDV	√	√	X	X	X
	Maximum CTD	√	√	X	X	√
	Minimum CLR	√	√	X	X	√
	Feedback	X	X	X	√	X

Note: : Specified; X: Unspecified.

12.3 Mechanisms for Traffic Management

Traffic management deals with the traffic control and congestion control procedures for ATM. ATM layer traffic control refers to the set of actions taken by the network to avoid congestion conditions. ATM layer congestion control refers to the set of actions taken by the network to minimize the intensity, spread, and duration of congestion.

When a connection is requested, the *connection admission control* (CAC) function of the switch determines, based on its available resources, whether an incoming connection request will be accepted or denied and ensures that existing connections are not affected when a new one is established. The CAC will only accept the new connection if it can meet its defined QoS parameters. Once an ATM connection is set up, a traffic contract specifying the negotiated characteristics of an ATM connection is agreed between the user and the network. The network commits itself to provide the type of guarantee appropriate to the service category, as long as the user keeps the traffic on the connection within the envelope defined by the traffic parameters.

A traffic contract specifies the negotiated characteristics of a VP/VC connection at an ATM UNI. The traffic contract at the public UNI consists of a connection traffic descriptor, a set of QoS parameters for each direction of the ATM layer connection, and the conformance definition. The values of the traffic contract parameters can be specified either explicitly or implicitly. A parameter value is explicitly specified in the initial call establishment message. This can be accomplished via signaling for switched virtual connections or via the NMS for permanent virtual connections at subscription time. A parameter value is implicitly specified when its value is assigned by the network using default rules.

The connection traffic descriptor consists of all parameters and the conformance definition used to specify unambiguously the conforming cells of the connection, including the following:

- Source traffic descriptor, which is a set of parameters that describes an inherent characteristic of an ATM traffic source;
- *Cell delay variance tolerance* (CDVT);
- Conformance definition.

A conformance definition is a theoretical description of how the traffic should behave to comply with the traffic descriptors. The conformance definition contains a sequence of *generic cell rate algorithms* (GCRA) [1] that are applied to each set of traffic descriptors. With the help of these algorithms, each cell of an ATM cell flow is either defined as conforming or nonconforming. A *leaky bucket* mechanism is an implementation of the algorithm. Each leaky bucket can be expressed as GCRA(I, L), which has two parameters:

- I: The increment parameter corresponds to the inverse of the compliant rate (fill rate of the bucket).
- L: The limit parameter corresponds to the number of cells that can burst at a higher rate (size of the bucket).

Figure 12.1 demonstrates the operation of a leaky bucket. Incoming cells flow into a "leaky bucket" which drains at a rate specified by I. The cells that drain from the bucket are referred to as conforming cells which are allowed to enter the network. If the cells arrive faster than they drain out of the bucket, the bucket will eventually overflow once a predefined level L is reached. The

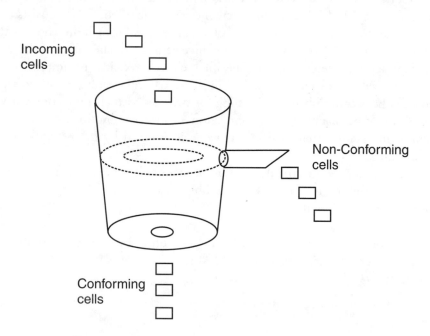

Figure 12.1 Example of a two-stage leaky bucket.

overflow cells are referred to as non-conforming cells, which are either discarded or tagged to compete for any unallocated resources.

In order to prevent a user from exceeding its traffic contract, which may affect the network resources assigned for another user, it is essential for the network to enforce the traffic contract. This is implemented by the *usage parameter control* (UPC) process at the UNI and optionally by a *network parameter control* (NPC) process at the NNI. These processes are better known as traffic policing. The UPC/NPC mechanism polices the traffic of the connection to detect non-conforming cells, and takes appropriate action on these cells to prevent them from affecting the QoS of the conforming cells of other connections. In this way, users are isolated from one another and are not affected by any "misbehaving" users.

Traffic shaping is a mechanism that alters the traffic characteristics of a stream of cells on a connection to achieve better network efficiency (while meeting the QoS objectives) or to ensure conformance at a subsequent interface. Usually, traffic shaping at the egress side of an ATM node is provided to protect subsequent ATM switches with small buffers from data bursts leading to buffer overflow. In addition, egress shaping is required to ensure that VPs originating at the ATM switch conform to the traffic contract. However, this mechanism can also be used at the ingress side to modify the traffic flow of terminal equipment and ATM switches that are not able to keep the traffic contract and are, for example, sending ATM cells with a higher peak bit rate than allowed. Traffic shaping adapts the traffic characteristics of the ATM cell stream to the traffic contract. This leads to a more efficient use of network resources by allowing network operators to allocate only those resources required according to the traffic contract.

When confronted by congestion, many ATM switches discard cells according to CLP. Considering that TCP/IP is used to transfer application data in which a PDU may consist of many ATM cells, the loss of a single cell could cause significant retransmissions, further aggravating congestion. Thus, discarding cells at frame level rather than at the cell level is more preferable. The mechanism by which a network device decides to drop frames is implementation-specific. Currently, there are two mechanisms that are used widely, namely *early packet discard* (EPD) and *trailing packet discard* (TPD) or *partial packet discard* (PPD).

EPD is a technique in which the switch starts to discard all cells except the last cell of an AAL5 PDU (identified by PT=1) from newly arriving packets when the switch buffer queues reach a user-configurable threshold level. If the first cell of a packet has entered the buffer, all remaining cells of the packet are also allowed to enter if sufficient buffer space is available.

TPD/PPD is an intelligent packet-discard mechanism that permits an ATM switch to hold down on retransmissions. When one or more of the cells in an AAL 5 frame are lost, the network marks subsequent cells so that a switch further down the line knows it can dump them, thus conserving bandwidth.

EPD and TPD/PPD can greatly reduce the number of retransmissions and thus significantly increase the "goodput," or effective throughput, by dropping off entire packets from a small number of connections rather than dropping cells from a large number of connections.

12.4   Traffic Model for ILMI 4.0 MIB

The traffic model for the ILMI 4.0 MIB [2] conforms to the TM4.0 specification [1], which is described by a set of traffic descriptors. Each of the traffic descriptors consists of three parts (see Figure 12.2) as follows:

- Type: Traffic descriptor type identifies the type of ATM traffic descriptor. The type may indicate no traffic descriptor or a traffic descriptor with one or more parameters specified as a parameter vector.

- Parameter vector: Containing five parameters, traffic descriptor parameter 1–5. The meanings of these parameters depend on the traffic descriptor type.

- Misc: Including information to further clarify the traffic being described, specifically the service category and the best effort indicator. The former categorizes the traffic type, which is one of the following: other, CBR, rtVBR, nrtVBR, ABR, or UBR. The latter indicates whether the traffic type is the "best effort" or not.

Five traffic descriptor types are specified in the ILMI 4.0 MIB. The mappings of the traffic types and their corresponding parameter definitions are listed in Table 12.2.

The above traffic descriptors are defined according to the conformance definitions in the "Traffic Management Specification 4.0 [1]." Each descriptor corresponds to a conformance definition and vice versa. The only exception is that the first row, the "No CLP/No SCR Type," can be used for three conformance definitions: CBR.1, UBR.1, and UBR.2. To differentiate between the CBR and the UBR categories, the best effort indicator is used. If this object is

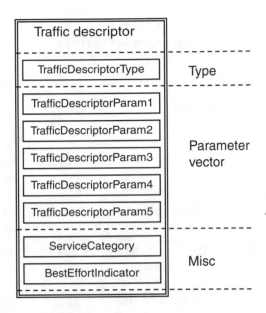

Figure 12.2 Traffic Descriptor for ILMI 4.0 MIB.

set to false, the network is requested to apply the CBR.1 conformance definition. If this object is set to true, the network is requested to apply the UBR.1 or
UBR.2 conformance definition [2]. However, it is not clear how to distinguish
between UBR.1 and UBR.2.

It should be noted that the traffic descriptors actually include all the
information of a traffic contract, instead of just the source traffic descriptor as
defined in TM4.0 [1].

12.5 Traffic Model for AToM MIB

The traffic descriptors for the AToM MIB have the same structure as those in
the ILMI 4.0 MIB; that is, each consists of three parts, type, parameter, vector
and other information as depicted in Figure 12.3. The last part contains only
one object, that is, the QoS Class. This is because the traffic model on which
AToM MIB is based is an earlier version defined in the ATM Forum's UNI
3.0/3.1 [3, 4] specifications, where four service classes were specified:

- Service class A: Constant bit rate video and circuit emulation;

Table 12.2
Traffic Descriptors for ILMI 4.0 MIB

Traffic Descriptor Type Name	Traffic Parameters					Confor-mance Supported
	1	2	3	4	5	
No CLP / No SCR Type	CLP=0+1 PCR	CDVT	Not used	Not used	Not used	CBR.1 UBR.1 UBR.2
SCR/No CLP Type	CLP=0+1 PCR	CLP=0+1 SCR	MBS	CDVT	Not used	VBR.1
CLP without tagging/ SCR Type	CLP=0+1 PCR	CLP=0 SCR	MBS	CDVT	Not used	VBR.2
CLP with tagging/ SCR type	CLP=0+1 PCR	CLP=0 SCR with excess traffic tagged as CLP=1	MBS	CDVT	Not used	VBR.3
CLP without tagging/MCR type	CLP=0+1 PCR	CDVT	MCR	Not used	Not used	AB

- Service class B: Variable bit rate video/audio;
- Service class C: Connection-oriented data;
- Service class D: Connectionless data.

Accordingly, four QoS classes numbered 1, 2, 3, and 4 have been specified with the aim of supporting service classes A, B, C, and D, respectively. An unspecified QoS Class numbered "0" is used for best effort traffic. Note that the AToM MIB is being adapted on this (new RFC superseding 1695 is expected).

Seven types of traffic have been specified in the AToM MIB. The mapping of the traffic types and their corresponding parameter definitions can be found in Table 12.3. In contrast to the ILMI 4.0 MIB, in which CDVT is specified for every traffic type, the AToM MIB implicitly specifies this parameter by the QoS class object.

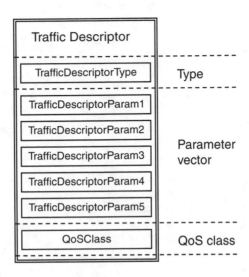

Figure 12.3 Traffic descriptor.

Table 12.3
Traffic Descriptors for AToM MIB

Traffic Descriptor Type Name	Traffic Parameters				
	1	2	3	4	5
None traffic descriptor type	Not used	Not used	Not used	Not used	Not used
No CLP/ No SCR type	CLP=0+1 PCR	Not used	Not used	Not used	Not used
CLP without tagging/ No SCR type	CLP=0+1 PCR	CLP=0 PCR	Not used	Not used	Not used
CLP with tagging/ No SCR type	CLP=0+1 PCR	CLP=0 PCR with excess traffic tagged as CLP=1	Not used	Not used	Not used
SCR/No CLP type	CLP=0+1 PCR	CLP=0+1 SCR	CLP=0+1 MBR	Not used	Not used

Table 12.3 (continued)

Traffic Descriptor Type Name	Traffic Parameters				
	1	2	3	4	5
CLP without tagging/ SCR type	CLP=0+1 PCR	CLP=0 SCR	CLP=0 MBR	Not used	Not used
CLP with tagging/ SCR type	CLP=0+1 PCR	CLP=0 SCR with excess traffic tagged as CLP=1	CLP=0 MBR	Not used	Not used

The above traffic descriptor definitions are based on the types of UPC functions defined in [3, 4], from which a user or a network manger can choose according to the traffic requirements of the connection to be established. There are three conformance definitions specified as follows:

- Conformance definition for PCR;

- Conformance definition for PCR CLP=0+1 and SCR CLP=0;

- Conformance definition for PCR CLP=0+1 and SCR CLP=0+1.

12.5.1 Conformance Definition for PCR

This definition specifies PCR for the CLP=0 cell stream and PCR for the CLP=0+1 cell stream, resulting in the following three traffic descriptor types:

- No CLP/no SCR type;

- CLP without tagging/no SCR type;

- CLP with tagging/no SCR type.

12.5.1.1 Conformance Definition

The conformance definition for PCR is defined as follows:

1. One GCRA(T_{0+1}, τ) defining the CDVT in relation to the PCR of the CLP=0+1 cell stream;

2. One GCRA(T_0, τ) defining the CDVT in relation to the PCR of the CLP=0 cell stream.

The "CLP with tagging/no SCR type" should be used if the user requests tagging and if tagging is supported by the network; otherwise, the choice should be the "CLP without tagging/no SCR type."

For networks that handle cells of the connection independent of the value of the CLP bit, the above conformance definition reduces to GCRA (1). The "No CLP/no SCR type" should be used, as the tagging option is not applicable.

12.5.2   Conformance Definition for PCR CLP=0+1 and SCR CLP=0

This definition specifies PCR for the CLP=0+1 cell stream and SCR for the CLP=0 cell stream, resulting in the following two traffic descriptor types:

- CLP without tagging/SCR type;
- CLP with tagging/SCR type

12.5.2.1   Conformance Definition:

The conformance definition for PCR CLP = 0+1 and SCR CLP = 0 is as follows:

1. One GCRA(T_{0+1}, τ) defining the CDVT in relation to the PCR of the CLP=0+1 cell stream;

2. One GCRA(T_{s0}, $\tau_{s0}+\tau$) defining the sum of the CDVT and the BT in relation to the SCR of the CLP=0 cell stream.

The "CLP with tagging/SCR type" should be used if the user requests tagging and if tagging is supported by the network; otherwise the "CLP with tagging/SCR type" should be used.

This conformance definition allows a connection to send CLP=1 cells at a PCR equal to the specified PCR of the CLP=0+1 cell stream.

12.5.3   Conformance Definition for PCR CLP=0+1 and SCR CLP=0+1

This definition specifies PCR for the CLP=0+1 cell stream and SCR for the CLP=0+1 cell stream, resulting in the following traffic descriptor type:

SCR/no CLP type

12.5.3.1   Conformance Definition

The conformance for PCR CLP = 0+1 and SCR CLR = 0+1is defined as:

1. One GCRA(T_{0+1}, τ) defining the CDVT in relation to the PCR of the CLP=0+1 cell stream;
2. One GCRA(T_{s0+1}, $\tau_{s0+1}+\tau$) defining the sum of the CDVT and the BT in relation to the SCR of the CLP=0+1 cell stream.

Since the tagging option is not applicable to this conformance definition, the only type for this conformance check is the "SCR/no CLP type."

Table 12.4 summarizes the traffic descriptor definitions in such a way that the rules for naming them are revealed. The traffic descriptor types are actually specified in terms of three options:

- Whether the SCR is checked;
- Whether CLP=0 cells are checked specifically;
- Whether excess CLP=0 cells are tagged with CLP=1.

A "0" in the table means "no" to an option, whereas a "1" corresponds to "yes." Note that the tagging option is not applicable if the CLP=0 cells are not checked specifically.

Possible mappings between TM 4.0 service categories and the UNI 3.1 QoS classes are shown in Table 12.5. The VBR in UNI 3.1 is, in fact, the rt-VBR in the TM 4.0, whereas the ABR and nrt-VBR classes were not defined in UNI 3.1. Both connection-oriented and connectionless data correspond to UBR in TM 4.0.

When setting up a connection, the requesting node informs the network of the type of service required, the traffic parameters of the data flows in each direction of the connection, and the QoS requested for each direction. Together, these form the traffic descriptors for the connection. In UNI 3.1, the QoS requested for each direction is not explicitly specified. Instead, the

Table 12.4
Summary of Traffic Descriptor Type Definitions

SCR	CLP	Tagging	Descriptor Type	OID Name*
0	0	X**	No CLP/No SCR type	NoClpNoScr
0	1	0	CLP without tagging/ No SCR type	ClpNoTaggingNoScr
0	1	1	CLP with tagging/ No SCR type	ClpTaggingNoScr
1	0	X**	SCR/No CLP type	NoClpScr
1	1	0	CLP without tagging/ SCR type	ClpNoTaggingScr
1	1	1	CLP with tagging/ SCR type	ClpTaggingScr

* The name does not include the prefix, which is "atm" for the AToM or "atmf" for ILMI MIB.
** Do not care bit.

Table 12.5
Mapping Between TM 4.0 Service Categories and UNI 3.1 QoS Classes

UNI 3.1 Service Class	TM 4.0 Service Category
A	CBR
B	rt-VBR
C	UBR
D	UBR
None	nrt-VBR
None	ABR

network offers a number of specified QoS classes that correspond to some or all of the QoS service types. The network administration has the responsibility of ensuring that the network is configured such that each of the offered QoS classes provides levels of QoS appropriate for each QoS type. The ATM Forum decided, however, that this method was too ambiguous and replaced it in UNI 4.0 [5] with explicit signaling of QoS parameters, desired values of which are requested at connection set-up time. UNI 4.0 signaling messages will carry both the QoS service classes and the explicit parameters, so that switches could operate on either, depending upon their own implementation.

12.6   LS1010 Traffic Management

In addition to the traffic management objects defined in the AToM MIB, the traffic management in LS1010 encompasses three Cisco MIB extensions [6]. The roles of each Cisco extension are described as follows:

- CISCO-ATM-TRAFFIC-MIB: Extends traffic model of the AToM MIB;
- CISCO-ATM-CONN-MIB: Configures operations and monitors the performances of virtual connections regarding traffic management;
- CISCO-ATM-RM-MIB: Manages switch resources.

12.6.1   Traffic Model

The traffic model in LS1010 is based on the UNI 3.0/3.1 specification, with Cisco proprietary extensions in order to support service categories that are specified in [1]. Figure 12.4 depicts the objects that are defined by Cisco in the CISCO-ATM-TRAFFIC-MIB. The "No CLP / No SCR CDVT Type" object is added into the traffic descriptor group in the AToM MIB. It specifies PCR for CLP0+1 and CDVT. This traffic descriptor is for CBR, ABR, or UBR service categories, which desire to specify the CDVT as well as PCR. This object is, in fact, the same as the "No CLP/no SCR type" in the ILMI 4.0 MIB.

In addition, two further parameters, the traffic-explicit service category and traffic-derived service category, are defined to provide additional information specifying service category for a VCC/VPC. These two objects facilitate an ATM switch to automatically choose default traffic contract parameters based on the service category requested.

The traffic explicit service category defines six values such as CBR, rt-VBR, nrt-VBR, ABR, UBR, and NotDefined, indicating the service category specified by the traffic descriptor. If it is defined, the setting of this object

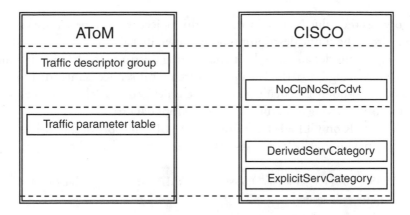

Figure 12.4 Cisco extension to the traffic model.

determines the service category used for the connection. This then limits the possible traffic descriptor types as listed in Table 12.6.

The traffic-derived service category object may be one of five types, CBR, rt-VBR, nrt-VBR, ABR, and UBR. This object indicates the service category derived from the traffic descriptor. If the traffic-explicit service category object has a defined value, then this object mirrors that value. If the explicit service

Table 12.6
CISCO ATM Explicit Service Categories and Their Associated Traffic Types

Explicit Service Category	Possible Type
CBR	NoClpNoScr
	ClpNoTaggingNoScr
	ClpTaggingNoScr
	NoClpNoScrCdvt
ABR/UBR	NoTrafficDescriptor
	NoClpNoScr.
	NoClpNoScrCdvt
VBR	NoClpScr
	ClpNoTaggingScr
	ClpTaggingScr

category has the value NotDefined, then this object reflects the service category derived from the traffic descriptor type as shown in Table 12.7.

Let us consider an example of configuring unstructured E1 circuit emulation services using LS 1010. The unstructured CBR service is intended to emulate a point-to-point E1 circuit. The service is defined as a "clear channel pipe," transparently carrying any arbitrary 2.048-Mbps data stream.

The PCR on CLP=0+1 required for AAL1 transport of 2,048-Kbps user data with maximum clock error +/- 50 ppm is calculated as follows:

$$2.048 \times 10^6 \, (1 + 50 \times 10^{-6}) \text{ bits/s} \, / \, (47 \text{ AAL1 octets/cell x 8 bits/octet}) = 5447.08 \text{ cells/second}$$

Thus, the PCR = 5448 cells per second. Assume that the CDVT for this switch is 30 ms. Table 12.8 lists traffic parameters to represent such a service by the AToM, ILMI 4.0 MIB, and the Cisco extension MIB, respectively.

12.6.2 Traffic Management for Virtual Connections

The objects that are defined in the CISCO-ATM-CONN-MIB for traffic management can be divided into two categories: configuration information and per-virtual connection statistics for cells dropped due to violations of traffic contracts. These objects are specified at both the VP level and the VC level, respectively, as depicted in Figure 12.5.

Table 12.7

Derived Service Category From Traffic Descriptor Type

Traffic Descriptor Type	Service Category
NoTrafficDescriptor	UBR
NoClpNoScr	CBR
ClpNoTaggingNoScr	
ClpTaggingNoScr	
NoClpNoScrCdvt	
NoClpScr	rt-VBR
ClpNoTaggingScr	
ClpTaggingScr	

Table 12.8
E1 CES Traffic Parameters for AToM MIB, ILMI 4.0 MIB, and Cisco MIB

| MIB | Traffic Parameters | | | |
	Type	PCR (Cells/s)	CDVT	Service Category/QoS Class
AToM	No CLP/No SCR *	5448	N/A	ClassA(1)
ILMI 4.0	No CLP/No SCR	5448	3000 (in 10 uS)	CBR(2)
Cisco	No CLP/No SCR CDVT*	5448	163 (in cell-times)	CBR(1) (for both the explicit and derived service category objects)
* Other types may be used.				

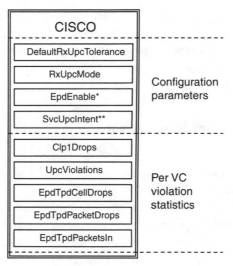

Figure 12.5 Cisco extensions for traffic management of virtual connections.

12.6.2.1 Configuration Information

The following objects are used to configure the traffic policing and the EPD operation for ATM virtual connections as well as for ATM interface.

- Default receive UPC tolerance: This object enables the use of user-specified tolerance parameters, an option not available in the AToM MIB. If the tolerance (i.e., CDVT/MBS) requested for UPC is not explicitly specified in virtual link creation, this object contains the default that is used. Otherwise, the tolerance can be found in the traffic parameter table row specified by the receive traffic descriptor index;
- Receive UPC mode: UPC operation mode at this connection, which is pass, tagging, drop, or local shaping;
- EPD enable: Enabled or disabled EPD operation specified at this VC. Since EPD/TPD is based on AAL5, it is applicable only for VCC. Consequently, this object is for the VC level only.
- SVC UPC intent: Determines the policing action to take for switched or soft virtual connections established through this interface.

12.6.2.2   Per-Virtual Connection Violation Statistics

These objects collect per-virtual connection statistics such as number of dropped CLP=1 cells, number of completely received packets, number of dropped EPD/TPD packets, number of dropped EPD/TPD cells, and the number of UPC violations. This information is useful in monitoring the performance of traffic management.

12.6.3   Switch Resource Management

Cisco ATM switch resource management MIB is designed to manage switch resources including shared memory cell buffers and ATM interfaces. The structure of the MIB is depicted in Figure 12.6.

Two groups in the MIB are specified to manage an LS1010 ATM switch as a whole. The switch configuration group provides general resource management configuration information, such as the ABR feedback control mechanisms to be used. In addition, the default UNI QoS objectives for service categories such as CBR, rt-VBR, and nrt-VBR are also configurable via this group. The objectives include the maximum CTD, peak-to-peak CDV, CLRs for CLP=0, and CLP=0+1 cells.

The switch shared memory group is designed to manage resources for the shared memory architecture used in LS1010. The switch supports four different delay classes for service, each with two loss priority classes. The class-based priority scheduling algorithm can be used to reduce delay for the guaranteed services of real-time traffic. This group maintains a configurable limit on the

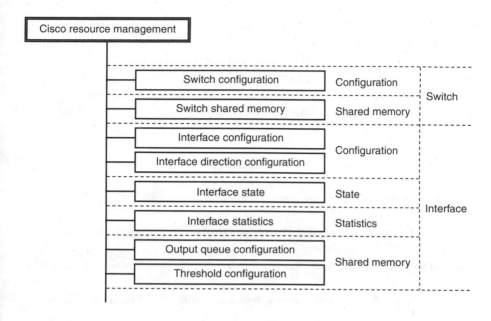

Figure 12.6 Structure of Cisco resource management MIB.

number of cells on all output queues of the switch and a current count of the number of cells for each priority level.

The remaining six groups are used to manage the resources for ATM interfaces. The configuration group, as its name suggests, provides information for interface configuration in relation to resource management. The group consists of two tables: The interface configuration table contains resource management configuration information for each ATM interface, including the default tolerances for each service category, while the interface direction configuration table holds configuration information for a traffic direction on an ATM interface, such as the maximum PCR and tolerance per service category.

The interface state group, on the other hand, contains per-service-category resource usage information, including the available cell rate and allocated cell rate for each direction and the estimated QoS performances for interfaces.

The interface statistics group collects statistics on resource allocation requests for each interface, which employs a number of counters for successful and failed requests due to various reasons.

Finally, the interface shared memory group is designed to manage resources for ATM interfaces using shared memory architecture—i.e., the output queues. Again, there are four queues for each interface, corresponding to four priority levels supported at the switch. This group comprises two tables, the output queue configuration table and the threshold configuration table. The former provides information about the output queues including queue length and configured and maximum sizes. The latter keeps the threshold levels for starting CLP/EPD and ABR flow control.

References

[1]   ATM Forum, "Traffic Management Specification Version 4.0," af-tm-0056.000, April 1996.

[2]   ATM Forum, "Integrated Local Management Interface (ILMI) Specification Version 4.0," af-ilmi-0065.000, September 1996.

[3]   ATM Forum, "ATM User-Network Interface Specification Version 3.0," September 1993.

[4]   ATM Forum, "ATM User-Network Interface Specification, Version 3.1," September 1994.

[5]   ATM Forum, "ATM User-Network Interface Signaling Specification Version 4.0," af-sig-0061.000, July 1996.

[6]   Cisco LightStream 1010 ATM Switch, Cisco System Inc., http://www.cisco.com/warp/public/730/LS1010.

13

OAM Flow Support

13.1 Introduction

The OAM functionality defined in the ITU-T I.610 [1], as introduced in Chapter 4, provides mechanisms for fault and performance management of ATM networks. LS1010 implements a subset of the ITU-T I.610 [1] specified in the ATM Forum's UNI 3.0 / 3.1 [2, 3] standards, which is focused on fault management, particularly alarm surveillance and connectivity verification functions.

Similar to many other ATM standards, the OAM specifications remain in a state of evolution. Due to the fact that some of the details have not been fully specified, there are a number of issues that are subject to vendor differentiation, which, in turn, may cause inter-operability problems in the future.

After considerable delay, the "first cut" implementations of the standards defined for OAM are now appearing in ATM switches. However, the level of implementation, integration, and reliability of OAM is variable. The LS1010 is amongst the early implementations, supporting F4 and F5 ATM OAM segment and end-to-end flows periodically or on demand. Furthermore, the LS1010 incorporates unique value-added OAM Ping functions that are built upon ATM OAM cells, enabling the easy monitoring of VP/VC level connection and link status [4].

13.2 OAM Cell Formats

The ATM Layer contains two highest OAM levels, the F4 and the F5 OAM flows. These flows can be classified into two types described as follows:

- End-to-end: Flows between ATM connection end-points;
- Segment: Flows across one or more interconnected concatenations of virtual links.

The ITU-T I.610 [1] defines the general formats for VP/VC level OAM cells, whose format is detailed in [3] and depicted in Figure 13.1. The payload of an OAM cell consists of the following fields:

- OAM cell type: Four-bit indicator of the type of OAM cell. Currently, only one type, 0001, is defined by the UNI3.1 to indicate fault management.
- Function type: Four-bit indicator gives the purpose of this particular OAM cell:
  - 0000: AIS;
  - 0001: *Receive defect indication* (RDI), formerly known as FERF;
  - 1000: Loopback.
- Function-specific Fields: 45-byte specifying the functions and information for this particular cell and denoting destination and failure information.

F4 (VPC) OAM Cell

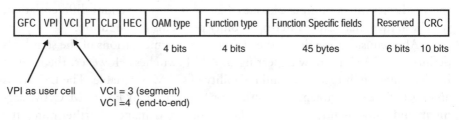

F5 (VCC) OAM Cell

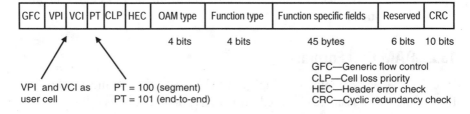

Figure 13.1 Common OAM cell format.

- Reserved: Six-bit of pad reserved for future use.
- CRC-10: 10-bit *cyclical redundancy check* (CRC) for the previous 46 bytes and the reserved field.

The VPC operational information is carried via the F4 flow OAM cells. These cells have the same VPI value as the user-date cells. Their types are identified by two preassigned VCI values. The VCI = 3 is used to identify a segment OAM flow, while VCI = 4 identifies an end-to-end flow.

The format of F5 OAM cells is slightly different from that of F4 cells, in that the PTI field is used to indicate whether an OAM cell flows between a segment or the whole span of a virtual connection. PTI = 100 specifies a segment F5 flow. PTI=101, on the other hand, represents an end-to-end FS flow.

Note that OAM cells do not form part of the user information stream transferred to or from higher layer protocols or user applications.

13.2.1 Alarm Surveillance

The alarm surveillance cells are used to indicate the failure of a virtual connection. When a failure is detected, OAM AIS cells are inserted in all the downstream VPCs/VCCs affected by this failure. At the terminating point of the affected VPCs/VCCs, the error status is reflected back upstream by inserting OAM RDI cells. Please see Chapter 4 for details.

The function-specific fields for alarm surveillance cells are depicted in Figure 13.2, which contains failure type and failure location information. However, the detailed coding of these fields has not yet been standardized.

13.2.2 Connectivity Verification

In contrast to physical connections where a connection has to be taken out of service when loopback tests are performed, I.610 defines two mechanisms, namely the continuity check and the loopback test, that allow nonintrusive

Failure type	Failure location	Unused
1 byte	16 bytes	28 bytes

Figure 13.2 Function-specific fields for AIS/RDI cells.

tests of connection integrity. However, only one mechanism has been standardized for connectivity verification in the UNI 3.0/3.1 [2, 3], that is, loopback. The function-specific fields for alarm surveillance cells are depicted in Figure 13.3 as explained below:

- Loopback indication: Indicates whether incoming cell is to be looped back. Only the least-significant bit is used. This bit is set by the sending endpoint and reset by the loopback destination.

- Correlation tag: Used to associate loopback cells sent with those received, e.g., a sequence number or a time stamp.

- Loopback location (optional): Identifies the point along a virtual connection where the loopback is to occur.

- Source identifier (optional): Identifies the originator of the loopback cell so as to identify returning loopback cells for collection and removal from the connection.

Due to the lack of a format specification, the loopback location and source identifier are again open to different interpretations by implementors. For example, these addresses may be designated as a portion of an ATM NSAP address. Without a common definition of these addresses, interoperable loopback testing is confined to endpoints, and the diagnostic potential of being able to test connectivity with intermediate nodes is lost.

13.3   OAM Support in LS1010

The Cisco LS1010 ATM switch supports F4 and F5 OAM cell flows, including segment and end-to-end for both VPCs and VCCs. These flows can be sent periodically or on demand. OAM functions also provide AIS and RDI support. All these management functions are based on two Cisco MIBs, the

Loopback indication	Correlation Tag	Loopback location Id	Source Id	Unused
1 byte	4 bytes	16 bytes	16 bytes	8 bytes

Figure 13.3 Function-specific fields for loopback cells (loopback indication more).

CISCO-ATM-CONN-MIB and the CISCO-OAM MIB. The latter will be detailed in Section 13.4.

The CISCO-ATM-CONN-MIB contains five objects for configuring OAM functions, depicted in Figure 13.4. Two of them, the AIS enable and RDI enable, are used to activate and deactivate the generation of AIS/RDI signals.

The remaining three objects provide information for the OAM loopback test. LS1010 supports two types of OAM loopback, the segment loopback and the end-to-end loopback, which are shown in Figures 13.5 and 13.6,

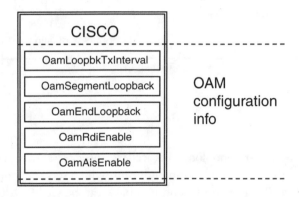

Figure 13.4 OAM configuration objects.

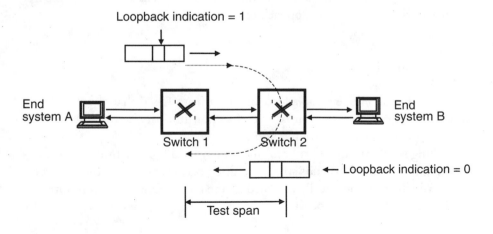

Figure 13.5 Segment loopback test.

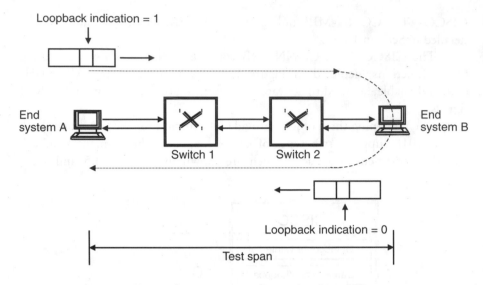

Figure 13.6 End-to-end loopback test.

respectively. An end-to-end loopback cell is inserted at one endpoint and looped back by the other connection endpoint, which sets the loopback indication from "1" to "0." The end-to-end loopback can be enabled/disabled by the end loopback object. Similarly, the segment loopback is performed between two preassigned segment loopback endpoints, which is configured via the segment loopback object. The loopback transmit interval is used to control the frequency of the generation of loopback cells.

13.4   OAM Ping

OAM Ping is a unique value-added function from Cisco that allows the network administrator to verify link or connection integrity at any intermediate point facilitating network troubleshooting.

Ping is perhaps the most basic and widely used tool for Internet management. When invoked, it sends small, serialized packets of data to the host computer, which are automatically returned by the remote computer. It is normally used for the following purposes:

- To determine whether a connection can be made to a remote host;
- To measure round-trip delay;

- To analyze the quality of the connection.

The functionality of the OAM Ping is very similar to that of ordinary Ping. However, their implementations are different. The latter is based on the ICMP echo request, whereas the former uses the OAM loopback protocol. Specifically, the OAM Ping makes use of OAM loopback cells with the loopback location ID field set to the IP or ATM address that corresponds to a specific node in a given network. When the loopback cell reaches the corresponding node, that node recognizes the loopback location ID as its own and loops it back. The source ID filed, on the other hand, contains the IP or ATM address representing the node that initiates the test. The initiating node checks this field from the OAM cells that are looped back to identify the cells sent by itself. Figure 13.7 depicts the operation of this type of loopback. Although OAM Ping can be supported by both end-to-end and segment flows, it is recommended that the former be used, because the OAM cells may be looped back by a segment endpoint before reaching their desired destination.

Unlike Internet Ping, which is usually initiated by a user directly, the OAM Ping function is invoked by the network manager via an SNMP protocol. A Cisco proprietary MIB, the CISCO-OAM-MIB, is specified for this purpose. Its structure is depicted in Figure 13.8.

A management station wishing to create an entry should first generate a pseudo-random serial number to be used as the index to this sparse table. The

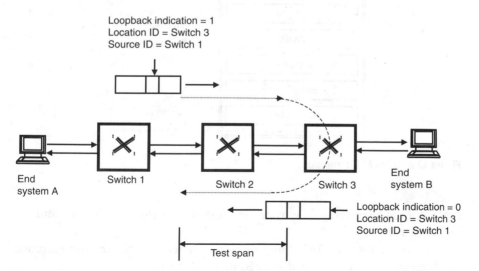

Figure 13.7 Loopback test using location identifier.

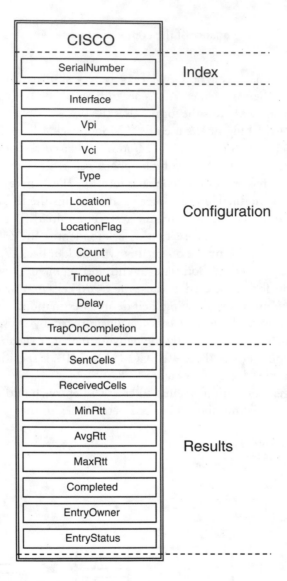

Figure 13.8 OAM MIB structure.

NMS should then create the associated instance of the entry owner and the entry status objects.

In order for the OAM Ping to work properly, necessary configuration information has to be supplied as listed in Table 13.1.

Table 13.1
OAM Ping Configuration Information

Type	Object(s)	Description
Virtual connection identification	Interface	ifIndex and VPI for F4 OAM flow or ifIndex, VPI, and VCI for F5 OAM flow
	VPI	
	VCI	
OAM cell setting	Type	End-to-end or segment
	Location	IP address or NSAP prefix of the node at which the OAM loopback is to occur
	Location flag	Type of address embedded into the location string (i.e., IP address, NSAP prefix, or a fixed value)
Timers and counters	Count	The number of loopback cells to be sent
	Delay	The amount of time to wait before sending the next OAM loopback cell in a sequence
	Timeout	The amount of time to wait for a response to transmitted OAM loopback cells before declaring the OAM loopback cells as "dropped"

Once the appropriate instance of all the configuration objects has been created, either by an explicit SNMP set request or by default, the entry status should be set to active to initiate the request.

When the sequence completes, the management station should retrieve the values of the status objects of interest and should then delete the entry. The MIB collects a number of statistics as follows:

- Number of cells sent;
- Number of cells received;
- Minimum round-trip time (Min RTT);
- Average round-trip time (Avg RTT);
- Maximum round-trip time (Max RTT).

The lost OAM cells can be found by comparing the number of cells sent against the number of cells received. If the numbers are not equal, some cells must be lost.

References

[1]   ITU-T Recommendation I.610, "B-ISDN Operation and Maintenance Principles and Functions," March 1993.

[2]   ATM Forum, "ATM User-Network Interface Specification, Version 3.0," September 1993.

[3]   ATM Forum, "ATM User-Network Interface Specification, Version 3.1," September 1994.

[4]   Cisco LightStream 1010 ATM Switch, Cisco System Inc., http://www.cisco.com/warp/public/730/LS1010.

14

Latest Developments

14.1 Limitations of the Current SNMP-Based ATM Management

SNMP is currently the dominating framework for managing ATM networks, through which the majority of ATM devices such as switches, routers, and workstations are now managed. It provides an integrated solution in which terminals, networks, and applications running over ATM networks can be managed within the same management framework.

A plethora of ATM-specific SNMP MIBs have been standardized; their roles range from ATM interface management, to virtual connection management, to application management. These MIBs, along with their vendor-specific counterparts, provide an unprecedented level of manageability for ATM networks.

However, SNMP-based ATM management does have its limitations. The first limitations come from the current SNMPv2 framework. As mentioned in Chapter 2, the SNMPv2 only specifies features that are necessary for device management. It does not support features that are considered important for network management, including proxy, informs, and remote configuration of the security and administration MIB objects. These are important issues when managing large networks such as ATM networks but were considered unnecessary overhead for a minimal conforming implementation.

Second, the AToM MIB does not fully incorporate the requirements of current ATM networks. The AToM MIB was developed when ATM was in its infancy. Consequently, the scope of the MIB is largely limited to permanent virtual connection management. It is essential that ATM management standards reflect recently completed major internetworking standards such as UNI 4.0, PNNI, and traffic management 4.0. In particular, the management standards should support the new service categories and new connection types, such as switched virtual connections, soft permanent virtual connections, and VP-tunneling.

Finally, the flat structure of the managed information within the SNMP framework may inherently prevent this approach from being scaled to very large networks, although it is not necessarily a disadvantage in managing a small- to medium-size network.

As of this writing, there are a lot of efforts under way that will help to tackle some aspects of the above limitations. Some of them are mature enough to become or have just been published as standards, whereas others are still in different phases of development. The standard-ready efforts will be briefly introduced in the following sections.

14.2   SNMP Version 3

In January 1998, SNMPv3 framework was published [1] as a proposed standard in a set of RFCs as listed in Table 14.1.

These specifications are built on the SNMPv2 draft standard protocol (RFC 1902-1908) with the addition of security, proxy, and remote configuration capabilities to SNMPv2. The objectives are the following.

Table 14.1
SNMP v3 RFCs

RFC No	Description
RFC 2271	An Architecture for Describing SNMP Management Frameworks
RFC 2272	Message Processing and Dispatching for the Simple Network Management Protocol (SNMP)
RFC 2273	SNMPv3 Applications
RFC 2274	User-Based Security Model (USM) for Version 3 of the Simple Network Management Protocol (SNMPv3)
RFC 2275	View-Based Access Control Model (VACM) for the Simple Network Management Protocol (SNMP)

- To accommodate the wide range of operational environments with differing management demands;
- To facilitate the need to transition from previous, multiple protocols to SNMPv3;
- To facilitate the ease of set up and maintenance activities.

In order to meet the above objectives, SNMPv3 adopts a modular architecture [2] as depicted in Figure 14.1. This approach makes it possible to incrementally upgrade portions of SNMP, without disrupting an entire SNMP framework. Each SNMPv3 entity consists of an SNMP engine and one or more associated applications.

An SNMP engine provides services for sending and receiving messages, authenticating and encrypting messages, and controlling access to managed objects. Accordingly, the engine contains a dispatcher, a message processing subsystem, a security subsystem, and an access control subsystem. In contrast to earlier versions of SNMP, the applications are not limited to run on a manager; an agent could be described as executing specific applications, such that all local processing is just a special case of an application. All traditional processing can be broken down to five types of applications—request generation, response processing, notification sending, notification receiving, and proxy processing.

The deployment of SNMPv3 will provide secure access to management information, which, in turn, will make it possible not only to monitor

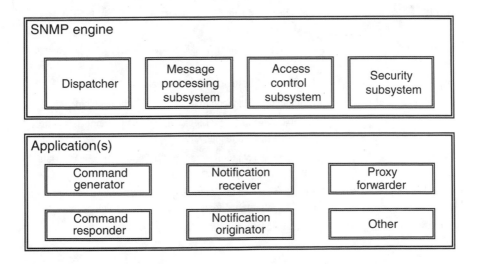

Figure 14.1 Architecture for SNMP v3 entity.

networks, but also to control them. It should be noted that the MIBs that have been covered in this book are *protocol-independent*— i.e., they work with all versions of SNMP.

14.3   IETF AToM Working Group

Since the AToM MIB was published in 1994, further work has been carried on by the IETF AToM Working Group to reflect growing experience and industry requirements for managing ATM networks. As a result, a number of Internet-drafts have been published as listed in Table 14.2. Some of the drafts extend the scope of SNMP-based ATM management into the areas that have not yet been addressed, while others refine existing standards. A number of vendors have already implemented some of the MIBs directly in their ATM switches. For example, the accounting MIBs are implemented in LS1010. These MIBs also have great influences on the design of vendor-specific proprietary MIBs. It is expected that these drafts will become Internet standards soon.

Table 14.2
Internet Drafts From the AToM Working Group

MIB Name	Description
ATM TC	Definitions of Textual Conventions and OBJECT-IDENTITIES for ATM Management
Supplemental	Definitions of Supplemental Managed Objects for ATM Management
Test	Definitions of Tests for ATM Management
Next iteration AToM	Definitions of Managed Objects for ATM Management
History TC	Textual Conventions for MIB Modules Using Performance History Based on 15-Minute Intervals
ATM History*	Managed Objects for Recording ATM Performance History Data Based on 15-Minute Intervals
Accounting Control	Managed Objects for Controlling the Collection and Storage of Accounting Information for Connection-Oriented Networks
Accounting Information	Accounting Information for ATM Networks

* It is unclear whether this one will be standardized.

The ATM TC MIB [3] defines a set of textual conventions available to ATM MIB modules for managing ATM-based interfaces, devices, networks and services.

The "ATM Supplemental MIB" [4] offers the functionality needed for the management of ATM switched virtual connections. Moreover, this MIB provides additional information to manage permanent virtual connections, soft virtual permanent connections, and VP tunneling.

The ATM Test MIB [5] specifies four scenarios of loopback tests over ATM virtual connections by making use of the ATM OAM F4/F5 loopback functions. The scenarios include segment, end-to-end, loopback tests using location identifier, and network loopback test. The first three are described in Chapter 13. This MIB utilizes a new testTable specified in the system/interfaces test MIB [6], which is also an Internet draft.

The next iteration AToM MIB [7] updates the traffic model to support traffic management 4.0. There are also other refinements of the existing AToM MIB.

The role of the ATM history MIB [8] is to specify managed objects to record and retrieve ATM performance history data recorded in 15-minute intervals, which is intended to satisfy the requirements defined by the ATM Forum in [9]. The history TC MIB defines the set of textual conventions to be used when performance history based on 15-minute intervals is kept. This is a mechanism used in the trunk MIBs, as described in Chapter 7. ATM history MIB uses history TC MIB [10].

The last two MIBs perform accounting management—i.e., collecting, interpreting, and reporting costing and charging oriented information on service usage. A common set of parameters and algorithms for basic support and interoperability is standardized by the MIBs as the basis for usage metering, charging, and billing of ATM networks and service provision. The accounting information MIB [11] defines a set of ATM-specific usage metering objects that can be collected for connections on ATM networks. The accounting control MIB [12] specifies managed objects, in a manner independent of the type of network, for controlling the selection, collection, and storage of usage metering objects into files for later retrieval via a file transfer protocol

14.4 Conclusion

Manageability is one of the keys to the success of ATM technology. SNMP is an well-understood network management framework, upon which a lot of efforts have been devoted to managing ATM networks. As a result, various ATM-specific MIB modules have been standardized, covering almost every area of ATM applications. Most of the ATM switches today are managed by SNMP. The SNMP-based approach features advantages such as low implementation cost and minimal hardware/software resources required in operation. Most importantly, with SNMP-based ATM management, it is possible to integrate network management, device management, and application management within the same management framework, which will enable flexible, interoperable, and cost-effective enterprise management. Moreover, SNMP is evolving toward becoming more powerful and effective. Accordingly, it is most likely that SNMP will remain the major framework for managing the enterprise/access ATM network at least in the near future.

References

[1] Harrington, D., "The Evolution of Architectural Concepts in the SNMPv3 Working Group," *The Simple Times,* Volume 5, Number 1, December 1997.

[2] Harrington, D., R. Presuhn, and B. Wijnen, "An Architecture for Describing SNMP Management Frameworks," RFC 2271, January 1998.

[3] Noto, M., and K. Tesink, "Definitions of Textual Conventions and OBJECT-IDENTITIES for ATM Management," Internet-Draft, January 1997, <draft-ietf-AToM MIB-atm2TC-06.txt>.

[4] Ly, F., M. Noto, A. Smith, E. M. Spiegel, and K. Tesink, "Definitions of Supplemental Managed Objects for ATM Management," Internet-Draft, July 1997, <draft-ietf-AToM MIB-atm2-11.txt>.

[5] Noto, M., and K. Tesink, "Definitions of Tests for ATM Management," Internet-Draft, June 1997, <draft-ietf-AToM MIB-test-03.txt>.

[6] McCloghrie, K., M. Greene, and K. Tesink, "Definitions of Managed Objects for System and Interface Testing," Internet-Draft, March 1997, <draft-ietf-ifmib-testmib-03.txt>.

[7] Tesink, K., "Definitions of Managed Objects for ATM Management," Internet-Draft, January 1997, <draft-ietf-AToM MIB-atm1ng-03.txt>.

[8] Mouradian, G., "Managed Objects for Recording ATM Performance History Data Based on 15 Minute Intervals," Internet-Draft, November 1996, <draft-ietf-AToM MIB-atmhist-00.txt>.

[9] ATM Forum, "M4 Interface Requirements and Logical MIB," af-nm-0020.000, 1994.

[10] Tesink, K., "Textual Conventions for MIB Modules Using Performance History Based on 15 Minute Intervals," October 1996, <draft-ietf-AToM MIB-perfhistTC-01.txt>.

[11] McCloghrie, K., J. Heinanen, W. Greene, and A. Prasad, "Accounting Information for ATM Networks," Internet-Draft, November 1996, <draft-ietf-AToM MIB-atmacct-01.txt>.

[12] McCloghrie, K., J. Heinanen, W. Greene, and A. Prasad, "Managed Objects for Controlling the Collection and Storage of Accounting Information for Connection-Oriented Networks," Internet-Draft, November 1996, <draft-ietf-AToM MIB-acct-04.txt>.

Appendix A
Current IETF Network Management RFCs

This appendix lists the IETF network management RFCs at the time of writing. For the most recent list, please check *The Simple Times* at:

http://www.simple-times.org/

A.1 SNMPv1 Framework

The *full standards* are listed as follows:

- RFC 1155: Structure of Management Information (SMI);
- RFC 1157: Simple Network Management Protocol (SNMP);
- RFC 1212: Concise MIB definitions.

The *proposed standards* are the following:

- RFC 1418L: SNMP Over OSI;
- RFC 1419: SNMP Over AppleTalk;
- RFC 1420: SNMP Over IPX.

The informational standard is

RFC 1215: A Convention for Defining Traps for Use With the SNMP

A.2 SNMPv2 Framework

The *draft standards* are the following:

- RFC 1902: SMI for SNMPv2;
- RFC 1903: Textual Conventions for SNMPv2;
- RFC 1904: Conformance Statements for SNMPv2;
- RFC 1905: Protocol Operations for SNMPv2;
- RFC 1906: Transport Mappings for SNMPv2;
- RFC 1907: MIB for SNMPv2;
- RFC 1908: Coexistence Between SNMPv1 and SNMPv2.
- RFC 1901: Introduction to Community-Based SNMPv2;
- RFC 1909: An Administrative Infrastructure for SNMPv2;
- RFC 1910: User-Based Security Model for SNMPv2.

A.3 SNMPv3

The *proposed standards* are the following:

- RFC 2275: View-Based Access Control Model (VACM) for SNMP;
- RFC 2274: User-Based Security Model (USM) for SNMPv3;
- RFC 2273: SNMPv3 Applications;
- RFC 2272: Message Processing and Dispatching for the SNMP;
- RFC 2271: An Architecture for Describing SNMP Management Frameworks.

A.4 MIB Modules

The *full standards* are listed as follows:

- RFC 1213: Management Information Base (MIB-II);
- RFC 1643: Ether-Like Interface Type (SNMPv1).

The *draft standards* are the following:

- RFC 1493: Bridge MIB;
- RFC 1559: DECnet phase IV MIB;
- RFC 1657: BGP version 4 MIB;
- RFC 1658: Character Device MIB;
- RFC 1659: RS-232 Interface Type MIB;
- RFC 1660: Parallel Printer Interface Type MIB;
- RFC 1694: SMDS Interface Protocol (SIP) Interface Type MIB;
- RFC 1724: RIP version 2 MIB;
- RFC 1748: IEEE 802.5 Token Ring Interface Type MIB;
- RFC 1757: Remote Network Monitoring MIB;
- RFC 1850: OSPF Version 2 MIB;
- RFC 2115: Frame Relay DTE Interface Type MIB.

The *proposed standards* are the following:

- RFC 1285: FDDI Interface Type (SMT 6.2) MIB;
- RFC 1381: X.25 LAPB MIB;
- RFC 1382: X.25 PLP MIB;
- RFC 1406: DS1/E1 Interface Type MIB;
- RFC 1407: DS3/E3 Interface Type MIB;
- RFC 1414: Identification MIB;
- RFC 1461: Multiprotocol Interconnect over X.25 MIB;
- RFC 1471: PPP Link Control Protocol (LCP) MIB;
- RFC 1472: PPP Security Protocols MIB;
- RFC 1473: PPP IP Network Control Protocol MIB;
- RFC 1474: PPP Bridge Network Control Protocol MIB;
- RFC 1512: FDDI Interface Type (SMT 7.3) MIB;
- RFC 1513: Token Ring Extensions to RMON MIB;

- RFC 1514: Host Resources MIB;
- RFC 1525: Source Routing Bridge MIB;
- RFC 1565: Network Services Monitoring MIB;
- RFC 1566: Mail Monitoring MIB;
- RFC 1567: X.500 Directory Monitoring MIB;
- RFC 1573: Evolution of the Interfaces Group of MIB-II;
- RFC 1595: SONET/SDH Interface Type MIB;
- RFC 1604: Frame Relay Service MIB;
- RFC 1611: DNS Server MIB;
- RFC 1612: DNS Resolver MIB;
- RFC 1628: Uninterruptible Power Supply MIB;
- RFC 1650: Ether-Like Interface Type (SNMPv2);
- RFC 1666: SNA NAU MIB;
- RFC 1695: ATM MIB;
- RFC 1696: Modem MIB;
- RFC 1697: Relational Database Management System MIB;
- RFC 1742: AppleTalk MIB;
- RFC 1747: SNA DLC MIB;
- RFC 1749: 802.5 Station Source Routing MIB;
- RFC 1759: Printer MIB;
- RFC 2006: Mobile IP MIB;
- RFC 2011: SNMPv2 IP MIB;
- RFC 2012: SNMPv2 TCP MIB;
- RFC 2013: SNMPv2 UDP MIB;
- RFC 2020: IEEE 802.12 Interfaces MIB;
- RFC 2021: RMON-2 MIB;
- RFC 2024: Data Link Switching MIB;
- RFC 2037: Entity MIB;
- RFC 2051:APPC MIB;
- RFC 2074: RMON Protocol Identifier;
- RFC 2096: IP Forwarding Table MIB;
- RFC 2108: IEEE 802.3 Repeater MIB;

- RFC 2127: ISDN MIB;
- RFC 2128: Dial Control MIB;
- RFC 2155: APPN MIB;
- RFC 2206: Resource Reservation Protocol MIB;
- RFC 2213: Integrated Services MIB;
- RFC 2214: Integrated Services Guaranteed Service Extensions MIB;
- RFC 2232: DLUR MIB;
- RFC 2233: Interfaces Group MIB;
- RFC 2238: High Performance Routing MIB;
- RFC 2239: IEEE 802.3 Medium Attachment Unit (MAU) MIB;
- RFC 2266: IEEE 802.12 Repeater MIB.

Appendix B
Relevant Standards From the ATM Forum

This appendix lists the relevant network management standards in relation to this book from the ATM Forum at the time of writing. For the most recent list please check ATM Forum's web site at:

http://www.atmforum.com/atmforum/specs/

The Data Exchange Interface Working Group standard is:

af-dxi-0014.000   Data Exchange Interface version 1.0

The Integrated Layer Management Interface Working Group is:

af-ilmi-0065.000   ILMI 4.0

- af-lane-0021.000: LAN Emulation Over ATM 1.0;
- af-lane-0038.000: LAN Emulation Client Management Specification;
- af-lane-0050.000: LANE 1.0 Addendum;
- af-lane-0057.000: LANE Servers Management Spec v1.0;
- af-lane-0084.000: LANE v2.0 LUNI Interface;
- af-mpoa-0087.000: Multi-Protocol Over ATM Specification v1.0.

The Network Management Working Group standards are listed as follows:

- af-nm-0019.000: Customer Network Management (CNM) for ATM Public Network Service;
- af-nm-0020.000: M4 Interface Requirements and Logical MIB;
- af-nm-0027.000: CMIP Specification for the M4 Interface;
- af-nm-0058.000: M4 Public Network View;
- af-nm-0071.000: M4 "NE View";
- af-nm-0072.000: Circuit Emulation Service Interworking Requirements, Logical and CMIP MIB;
- af-nm-0073.000: M4 Network View CMIP MIB Spec v1.0;
- af-nm-0074.000: M4 Network View Requirements & Logical MIB Addendum;
- af-nm-test-0080.000: ATM Remote Monitoring SNMP MIB.

The PNNI Working Group standards are the following:

- af-pnni-0026.000: Interim Inter-Switch Signalling Protocol;
- af-pnni-0055.000: P-NNI V1.0;
- af-pnni-0066.000: PNNI 1.0 Addendum (soft PVC MIB);
- af-pnni-0075.000: PNNI ABR Addendum;
- af-pnni-0081.000: PNNI v1.0 Errata and PICs.

The UNI signaling and traffic management standards are the following:

- af-uni-0010.001: ATM User-Network Interface Specification V3.0;
- af-uni-0010.002: ATM User-Network Interface Specification V3.1;
- af-sig-0061.000: UNI Signaling 4.0;
- af-sig-0076.000: Signaling ABR Addendum;
- af-tm-0056.000: Traffic Management 4.0;
- af-tm-0077.000: Traffic Management ABR Addendum.

The Voice and Telephony over ATM Working Group standards are listed as follows:

- af-vtoa-0078.000: Circuit Emulation Service 2.0;
- af-vtoa-0083.000: Voice and Telephony Over ATM to the Desktop;
- af-vtoa-0085.000: (DBCES) Dynamic Bandwidth Utilization in 64 KBPS Time Slot Trunking Over ATM - Using CES;
- af-vtoa-0089.000: ATM Trunking Using AAL1 for Narrow Band Services v1.0.

Appendix C
Useful Tools and URLs

The Internet plays an important role in the development of the SNMP framework, as the latter was originally designed to manage the former. There are a lot of useful resources regarding SNMP and ATM scattered throughout the Internet. This appendix provides a brief road map of those resources that are relevant to this book. Specifically, useful tools, including MIB compilers and browsers, and useful Uniform/Universal Resource Locators (URLs) will be introduced. Please visit the site by following the "Link to Network Management Sites" at:

http://www.artech-house.com

C.1 MIB Compiler

The role of an MIB compiler is equivalent to that of a C compiler. It compiles SNMP MIB modules written in ASN.1 into a format suitable for network management applications and/or management agents. An MIB compiler can also be used to check an MIB module to determine if its syntax is valid in defining the module. In addition, some MIB compilers provide user-friendly GUI, presenting MIB modules visually to its users in an well-organized, intuitive and structured manner; which makes it very easy for the users to understand the structure of MIB modules and lookup managed object definitions. For example, The MG-Soft Corporation (http://www.mg-soft.si/) provides such an MIB compiler. With MG-Soft's compiler, a user can check the MIB structure and the attributes of MIB objects, for instances, the data type, access level, description etc., which enables the user to traverse the MIB tree at ease. This MIB compiler can be downloaded at:

http://www.mg-soft.si/download.html

C.2 MIB Browser

An MIB browser is a simple network management application that allows the users to browse the MIB sub tree implemented in managed devices, in particular to retrieve and modify the instances of MIB objects. MIB modules compiled by the MG-Soft's MIB compiler can be loaded into the browser. Thus, the values of the objects can be checked against their definitions. Again, an MIB browser is also available for downloading from:

> http://www.mg-soft.si/download.html

In fact, just like many other vendors, MG-Soft offers both the MIB browser and compiler in a package—the MG-SOFT MIB Browser Professional Edition with MIB Compiler. Similar packages from other companies can be found from the SNMP Frequently Asked Questions (FAQ), the URL of which is listed in Table A.1.

Readers are encouraged to install such an MIB compiler and/or browser when reading this book in order to benefit from the well-defined visual interface for a quick understanding of the MIB modules that are introduced in this book. This would help you to get rid of many of the difficulties when reading MIB modules in plain text format; and moreover, to obtain hands-on experiences on how SNMP works.

C.3 Useful URLs

All the standards introduced in this book, except those from ITU and ISO, are accessible through the Internet. For example, all the MIBs, RFCs, and ATM Forum's standards can be downloaded from the Internet. Due to limited space, these documents are not included in the book. As an alternative, their URLs are provided instead, which are listed in Table A.1 and correspond to the order of chapters.

Table A.1
Useful URLs

Chapter	URL(s)	Description
1	None	
2	http://www.ietf.org/	Home page for the IETF

Chapter	URL(s)	Description
	http://www.isi.edu/rfc-editor/	Searchable RFC database
	http://www.pantherdig.com/snmpfaq/index.html	SNMP FAQ, a comprehensive source of information for SNMP
	http://snmp.cs.utwente.nl/ietf/rfc/index.shtml	Network management RFCs indexed in different ways
	http://www.simple-times.org/pub/simple-times/issues/	Simple Times, very good overviews of latest developments in SNMP
	http://www.snmpinfo.com/siupdate.htm	SNMPInfo's home page, providing all RFC MIB modules that are directly compileable by MIB Compilers
3	http://www.itu.ch/	Home page for the ITU-T
	http://www.atmforum.com/	Home page for the ATM Forum
	http://cell-relay.indiana.edu/cell-relay/FAQ/ATM-FAQ/FAQ.html	Cell Relay FAQ, containing comprehensive information on ATM
4	http://www.atmforum.com/atmforum/specs/approved.html	Approved ATM Forum's Specifications
5	ftp://ftp.atmforum.com/pub/approved-specs/	FTP server for approved ATM Forum's specifications, including documents and MIBs
6, 7	http://www.ietf.org/html.charters/atommib-charter.html	Home page for the IETF AToM MIB Working Group
	http://www.snmpinfo.com/siupdate.htm,	Try to download RFC MIBs from this site if you are going to compile them

Table C.1 (continued)

Chapter	URL(s)	Description
8, 9	ftp://ftp.atmforum.com/pub/approved-specs/	FTP server for approved ATM Forum's Specifications, including documents and MIBs
10, 11, 12, 13	ftp://ftp.cisco.com/pub/mibs/	Proprietary and standard MIBs implemented by Cisco, usually directly compileable.
14	http://www.ietf.org/html.charters/atommib-charter.html	Home page for the IETF AToM MIB Working Group
	http://www.ietf.org/html.charters/snmpv3-charter.html	Home page for the IETF SNMP v3 Working Group

It is recommended that readers download the MIB modules from the SNMPInfo's or Cisco's site whenever possible if a MIB compiler is used, as they are directly compileable. These modules are usually clean-up versions of the original MIB definitions that may be flawed by syntax mistakes or ambiguities. As a result, when a new MIB is defined, the module(s) are extracted and, if needed, corrected and posted at the above sites for download.

About the Author

Heng Pan received B.S. and M.S. degrees from Southeast University, China, in 1984 and 1990, respectively. Before recently joining the OpenCon System Inc. in the United States, he had worked at the University of Surrey, the University of Oxford, and the University of Bradford in the United Kingdom, as well as the Southeast University in China. He was actively involved in a number of high profile projects sponsored by the EC or the Chinese national programs on ISDN and ATM networks. His current research interests include ATM internetworking and network management.

Index

Lexicographic ordering, 29, 44
LGN. *See* Logical group node
LI. *See* Length indication
Line overhead, 189–90
Link aggregation, 213
Link group
   virtual channel, 164–66
   virtual path, 162–63
Link management information
      base, 135–48
Link, private network-network interface,
      209–10, 229, 232
Link trap, 69–70
Local area network, 1, 116, 243–44, 252
LOF. *See* Loss of frame
Logical link, 208–10, 229
Logical node, 206–8, 211–13, 230
Logical port, 310, 320–21
LOH. *See* Line overhead
Loop-back capability, 114
LOP. *See* Loss of pointer
LOS. *See* Loss of signal
Loss of frame, 194
Loss of pointer, 194
Loss of signal, 194
Lossy process, 219
LS1010. *See* Cisco LightStream 1010
LUNI. *See* LAN emulation service, user
      network interface

MAC. *See* Media access control
Managed element, 11–12
Managed node, 58
Management agent, 11
Management information base, 6, 9–10,
      12–16, 25–26, 36–37, 53, 107–8
   AToM, 155–56, 220
   broadcast and unknown server, 286–89
   conceptual tables, 53–57
   emulated LANE, 266–80
   enterprise-specific, 38
   interworking modules, 70–71
   LANE, 252–54, 280–86
   LEC structure, 256–57
   LS1010, 296–306, 311–16
   modules, 112, 128

proprietary, 294–96
   standard, 37–38
   view, 39, 42
   See also Trunk management
      information base
Management information base, ILMI
   address registration, 148–54
   link management, 136–48
   modules, 135
Management information base, PNNI
   groups, 221–42
   SNMP standards, 220
   structure, 220–21
Management information base II, 57–70
Management interface, 106–7
Management model, 11–12
Management plane, 77, 84, 85, 87, 92
Management station, 11
Manager-agent model, 183
Manager-agent request-response
      interaction, 38–39
Manager-manager request-response
      interaction, 38–39
Map table, 232–34, 236, 277
Max-access clause, 22, 25, 35
Maximum burst size, 325
Maximum cell delay variation, 217
Maximum cell rate, 217
Maximum cell transfer delay, 216
Max-repetition, 46–47
MBS. *See* Maximum burst size
MCDV. *See* Maximum cell delay variation
MCR. *See* Minimum cell rate
MCTD. *See* Maximum cell transfer delay
MD. *See* Mediation device
Media access control, 244, 247, 250–52,
      262, 273, 283
Mediation device, 118, 120, 121
Mediation function, 121
Medium group, 199
Message mode, 93, 95
Meta-signaling, 77, 87
Metrics group, 222, 236–37
MF. *See* Mediation function
MIB. *See* Management information base
MID. *See* Multiplexing indication

The Artech House Telecommunications Library

Vinton G. Cerf, Series Editor

For further information on these and other Artech House titles, including previously considered out-of-print books now available through our In-Print-Forever™ (IPF™) program, contact:

Artech House
685 Canton Street
Norwood, MA 02062
781-769-9750
Fax: 781-769-6334
Telex: 951-659
email: artech@artech-house.com

Artech House
Portland House, Stag Place
London SW1E 5XA England
+44 (0) 171-973-8077
Fax: +44 (0) 171-630-0166
Telex: 951-659
email: artech-uk@artech-house.com

Find us on the World Wide Web at: www.artech-house.com